ENVIRONMENTAL
ECONOMICS

ENVIRONMENTAL ECONOMICS

CONCEPTS, METHODS, AND POLICIES

Dodo Thampapillai

OXFORD

UNIVERSITY PRESS

OXFORD

UNIVERSITY PRESS

253 Normanby Road, South Melbourne, Victoria 3205, Australia

Oxford University Press is a department of the University of Oxford.
It furthers the University's objective of excellence in research, scholarship,
and education by publishing worldwide in

Oxford New York

Auckland Bangkok Buenos Aires Cape Town Chennai
Dar es Salaam Delhi Hong Kong Istanbul Karachi Kolkata
Kuala Lumpur Madrid Melbourne Mexico City Mumbai Nairobi
São Paulo Shanghai Taipei Tokyo Toronto

OXFORD is a trade mark of Oxford University Press
in the UK and in certain other countries

National Library of Australia

Cataloguing-in-Publication data:

Thampapillai, Dodo J.

Environmental economics: concepts, methods, and policies.

Bibliography.
Includes index.
ISBN 0 19 553577 4.

1. Environmental policy—Economic aspects. 2 Economic
development—Environmental aspects. I. Title.

333.7

Edited by Frances Wade
Text and cover designed by Racheal Stines
Illustrated and typeset by Kerry Cooke
Printed by Bookpac Production Services, Singapore

To
Professor Thambapillai Jogaratnam

Professor of Agricultural Economics and former head of the Postgraduate Research Institute, Faculty of Agriculture, University of Sri Lanka, Peradeniya. Without his help I would never have ventured into economics.

Contents

List of Figures

List of Tables

Preface

In preparing this text I have tried, as far as possible, to maintain a common theme, namely that *nature is capital and its depreciation should be explicitly accommodated in economic analyses*. As indicated in chapter 1, this theme and the recognition that needs to be afforded to the natural environment are not new to economics. However, there were certain periods of misplaced emphasis on the singular pursuit of economic growth. If it were not for such misplaced emphasis, it is possible that many aspects of environmental economics could have become integrated into what is commonly described as mainstream economics.

The text begins and concludes with the question of natural resource scarcity. It is organised into four parts. The first part, consisting of two chapters, is a study of natural resource scarcity and the linkages between the environment and the economy. These linkages are examined by recourse to the first and second laws of thermodynamics. The implications of the environment–economy linkages are then considered within the frameworks of microeconomics in part 2 (chapters 3–9), and those of macroeconomics in part 3 (chapters 10–13). Part 4 (chapter 14) is a reassessment of the question of natural resource scarcity.

The text is also designed for the delivery of courses in Environmental Economics at an introductory level as well as at an advanced level. Chapters 2, 3, 4, 7, 9, 10, 11, and 12 could be used for an introductory course spanning a semester of 13 weeks and for those with very little background in economics. For such an introductory course some sections of chapters 7 and 12 could be omitted. The remaining chapters could be used as the basis for an advanced course in environmental economics, especially for those who had done the introductory course. Such delivery of material was tested at Macquarie University, Swedish University of Agricultural Sciences (Uppsala), Christian Albrechts University of Kiel (Germany), and the National University of Singapore.

Acknowledgments

I remain grateful to several friends and colleagues at various universities in Australia and abroad. I owe a special word of thanks to Hans-Erik Uhlin (Swedish University of Agricultural Sciences and the University of Gavle), Claus-Hennig Hanf (Christian Albrechts University of Kiel), Bo Ohlmer and Lars Drake (Swedish University of Agricultural Sciences), and Euston Quah, Lysa Hong, and Yong Mun Cheong (National University of Singapore) for giving me the opportunity to deliver courses in their respective universities. This contributed significantly to the refinement of the text.

Claus Hanf, Hans-Erik Uhlin, and Euston Quah have been my co-authors in refereed journal articles, the content of which appears in this text. Lars Drake and Bo Ohlmer co-taught courses with me in both Sweden and Australia and used parts of the manuscript for this purpose. There are of course several others who helped in reviewing various chapters. My mentor, Jack Sinden, has been a constant source of encouragement with invaluable comments. Tim Swanson and Matthias Ruth reviewed parts of the macroeconomics chapters and also invited me to deliver the material to their graduate classes in economics at Cambridge University and Boston University. Shandre Thangavelu helped me with addressing some difficult issues, and especially with the final chapter. Hans Andersson and Peter Frykblom offered me valuable feedback on different sections of the text.

My colleagues at the Graduate School of the Environment and the Division of Environmental and Life Sciences at Macquarie University need a very special word of thanks not only for their collegial support, but also for permitting environmental economics to be an important component of the curriculum.

Last, but never least, I remain ever grateful to my wife Gowrie and children Vinoli and Dilan for their affection and support, without which none of this would have been possible.

Part 1

The Environment and the Economy

The Environment in Economics

The natural environment is the core of any economy. Without a natural environment economic activity cannot be sustained. That is, environmental sustainability is a necessary condition for economic sustainability. In recognition of these facts, the object of this text is to show how economic theory and its applications, especially in terms of the directions taken during the past three decades, should be modified towards achieving sustainability. But it would be unwise to say outright that economics had relegated the environment to irrelevance. We begin this chapter by showing that economic theory, both classical and neoclassical, has always recognised the pivotal role of the environment. However, a sense of environmental complacency pervaded economic policy during the three decades from 1950 to 1980. This was mainly due to a misplaced singular emphasis on economic growth. This misplaced emphasis was prompted by empirical evidence that appeared to rationally negate the possibility of environmental limits. However, the pervasion of environmental problems among both the rich and the poor has reintroduced the need to attain and maintain a proper balance between economic activity and the environment. The attainment and maintenance of this balance is the primary focus of environmental economics.

The classicals and the neoclassicals

Many observers tend to associate the natural environment with classical economics rather than neoclassical economics. It is true that in all classical theories (Adam Smith 1776; Thomas Malthus 1798; David Ricardo 1817; and John Stuart Mill 1848), economic growth was explicitly constrained by environmental limits.

But it is not quite true to say that neoclassical economics was founded on premises that ignored the importance of the natural environment. For example consider what Alfred Marshall (1891), who is perhaps the founder of modern neoclassical economics and the influential Cambridge school, had to say: 'In a sense there are only two agents of production, *nature* and man. But on the other hand man is himself largely formed by his surroundings, in which *nature* plays a great part' and '... labour and capital of a country, acting on its *natural resources*, produce annually a certain net aggregate of commodities ...'[1]

These statements recognise *nature* and *natural resources* as the ultimate factors of production. The implication is that if an economy is not endowed with natural resources, then it cannot produce goods and services. This same conclusion was reached, nearly a hundred years after Marshall, by the Brundtland Commission (1987, p. 37): 'Environment and development are not separate challenges; they are inexorably linked.'

Although environmental economics became established as a discipline within economics only in the 1960s, several economists had written extensively during the earlier part of the twentieth century on environmental issues and problems within the framework of neoclassical economics. Some important developments in economics owe their origins to theoretical and empirical work on environmental issues. For example the theory of externalities, which was introduced by Marshall in 1890 and subsequently formalised in welfare economics by Pigou in 1920, was primarily due to the recognition of events such as pollution, which fall outside the market. Irving Fisher, writing in 1904 on the various definitions of 'capital', considered natural endowments such as lakes and rivers to be capital assets and used them to illustrate the difference between stocks and flows. The concept of user costs was developed by Hotelling (1931) to account for the requirements of future generations. This was then taken up by Keynes (1936) for the explanation of permanent income, more popularly known these days as sustainable income. The issue of resource conservation within neoclassical economics dates back to at least Jevons (1866), who expressed concern over the rapid depletion of Britain's coal reserves. To name a few, more rigorous analyses of conservation issues during the earlier half of the twentieth century can be found in Gray (1914), Schikele (1935), Ciriacy-Wantrup (1938), Bunce (1942), and Scott (1954). The valuation of environmental goods also employed neoclassical tools, and one of the earliest applications was by Hotelling (1949). One of the earliest concise reviews of environmental valuation spanning the developments of the nineteenth century and early to mid twentieth century can be found in Sinden (1967).

Why, then, are the neoclassicals accused of disregard for the natural environment? The main reason is that, soon after World War II, the emphasis was almost singularly on economic growth, and the growth economists (for example Harrod

1939; Domar 1946; Solow 1956; and Samuelson 1948) assumed that output was determined primarily by labour and capital. Their exclusion of the natural environment from the production function was premised on the further assumption that the environment and its services constituted a given constant. This resulted in a long procession of theories that excluded the environment and sought to explain economic growth in terms of savings, investment, capital accumulation, labour productivity, and the substitutability between labour and capital. However, this exclusion was shortlived. Natural resource scarcity became an important argument in the growth models of the 1970s. The chief architects of this development were Karl-Goran Mäler from the Stockholm School of Economics, Partha Dasgupta from Cambridge University, John Hartwick from Queens University in Canada, and Robert Solow from the Massachusetts Institute of Technology. But some difficulties still persist with these models, which are reviewed in the next section along with the classical and neoclassical models.

Box 1.1 Some early ideas in neoclassical economics

A common Cambridge response to new ideas in economics has been 'But it is all in Marshall.' For example in his *Principles* Marshall says: 'Man does not create things. He only rearranges matter.' This statement is indeed the First Law of Thermodynamics, which we will consider in the next chapter. Marshall's description of nature as capital can be interpreted as follows. If one were to disaggregate any commodity into its components until one could disaggregate no more, then the ultimate components would be those that come from nature. At about the same time as Marshall, across the Atlantic Irving Fisher laid the foundations of capital theory and attributed three properties to capital. These are that it is durable, that it provides a flow of services, and that it depreciates with use. However, Fisher used nature as the basis for this conceptualisation. If we do not interfere with nature it can remain intact; that is, durable. Nature provides a flow of services—the air we breathe, the water we drink and the soil we till. And when we use nature for its services, it depreciates. We call this depreciation 'environmental degradation'.

A brief review of growth models

Although Malthus is the better known among the classicals, Ricardo presented a framework that is perhaps intellectually superior to that of Malthus. Ricardo treated the economy like a large farm with a fixed input of land. The increase in population involved increasing the amount of land to be made available for production. However, this also meant that less arable marginal land was being progressively brought into production. Hence, if we consider a national production function with land as the only input, the function displays diminishing marginal returns due to the utilisation of marginal lands, and productivity tapers off.

This is illustrated in figure 1.1. OA is the production function that tapers off as the economy's reliance on land of inferior quality increases. The straight line OB defines the cost of production. As the quality of the land declines, the costs of producing output must increase. Ricardo referred to this as the wage bill and assumed it to be linear. Hence the vertical distance between the production function and the costs of production describes the surplus or profits that accrue to the economy. This surplus diminishes (or gets squeezed out) as the marginal productivity of the land diminishes. At some point, say *b*, this surplus reduces to zero and becomes negative from there on. The classical view then is that an economy starts from a state of stagnation (point O), and moves to another state of stagnation (point *b*) through a period of growth, since population increase renders land to be a limiting factor. From a more modern perspective, the natural environment (and not just merely land) is believed to be the limiting factor.

As indicated, the neoclassical production function regards land (environment) as a given constant, and hence the arguments in the production function are only labour and capital. The contention of the early neoclassicals was that the limitations of fixed land size can be overcome by capital accumulation and the substitutions between labour and capital. The Harrod–Domar growth model is one of these and

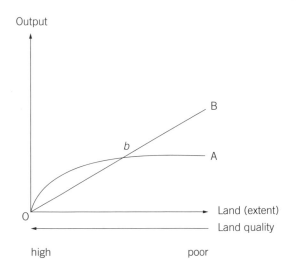

Figure 1.1   The Ricardian production function

can be briefly explained as follows. Suppose that the capital stocks and income during a given period are denoted by K and Y, and these in the following period are $(K + \Delta K)$ and $(Y + \Delta Y)$. If the capital–output ratio (κ), namely the amount of capital required to produce one unit of income, remains constant, then:

$$\kappa = K/Y = (K + \Delta K)/(Y + \Delta Y) = \Delta K/\Delta Y \qquad (1.1)$$

The change in capital stocks / Δ income = capital-output ratio

Hence it follows that the additions to capital stocks, which are also the savings (S), and which in turn become investment (I), can be defined as:

$$I = \Delta K = \kappa \Delta Y = S = \emptyset Y \qquad (1.2)$$

where ($\emptyset$) is the proportion of national income saved. The basic Harrod–Domar model is the result of dividing both sides of an identity in equation (1.2), expressed as ($\kappa \Delta Y = \emptyset Y$) by κY. Thus:

$$\Delta Y/Y = \emptyset/\kappa. \qquad (1.3)$$

The left-hand side of equation (1.3) is in fact the rate of economic growth, which is inversely proportional to the capital–output ratio and directly proportional to the savings ratio. In other words, the rate of economic growth will be high if capital is highly efficient (that is, if the capital–output ratio is low) and/or the savings ratio is high. Hence the strategies of economic growth that followed the models in the H–D vein emphasised the need to encourage savings and make investments in efficient forms of capital. Subsequently, the neoclassicals included technological change in their framework.

Figure 1.2 (p. 8) illustrates a basic Solow–Samuelson growth model that accommodates the role of technology. The left-hand panel explains the relationship between the return on capital (interest rate) and the amount of capital held per worker (capital–labour ratio). Assuming labour to be constant, it is evident that the interest rate (r) falls as capital accumulation occurs. That is, output response to increases in capital stocks follows the law of diminishing marginal returns. The right-hand panel shows the relationship between interest rates and real wages. As the interest rate falls, real wages increase (through a fall in the price level). Hence, in the absence of technology, it is possible to stimulate economic growth by encouraging capital accumulation, say from $(K/L)_0$ to $(K/L)_1$. The resulting fall in interest rates from r_0 to r_1 prompts an increase in real wages of (w_1-w_0). Technological improvements result in the two schedules of figure 1.2 being shifted outwards. As a result, real wages and capital stocks can increase simultaneously for a given interest rate.

A simple explanation of the growth models that recognised natural resource scarcity is provided in figure 1.3 (p. 9). The arguments in the production function are now labour, capital, and natural resources. The trade-off curves, labelled TT, describe

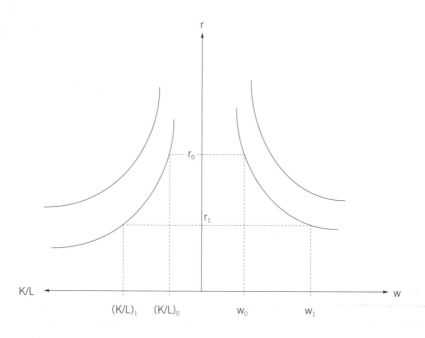

Figure 1.2 The neoclassical growth model

the substitutability between natural resources (R) and a composite labour–capital input (LK). As resource discoveries occur and/or the stock of renewable resources increases along with increases in LK, the trade-off curve shifts outwards. In figure 1.3A, these outward shifts are dominated by increases in the stocks of R, while in figure 1.3B they are dominated by increases in the stocks of LK.

The expansion (growth) path for the economy is given by the tangency of the output isoquants with the trade-off curves. The slope of the trade-off curve at these points of tangency describes the relative prices of R and LK. Although Dasgupta and Heal (1979) and Solow (1974) are often credited with demonstrating growth models of this type, the pioneer in this area was Mäler (1974), who was perhaps the first to exposit environmental economics within macroeconomic as well as microeconomic frameworks. In his macroeconomic analysis, Mäler formulated a model of economic growth that was constrained by the availability of natural resources. Some of the questions that were analysed subsequently, in the context of abstract theory, with these resource-constrained growth models were: Can aggregate consumption grow without bound for ever? What is the maximum level of consumption that can be maintained for ever?

Dasgupta and Heal (1979) illustrated that it is possible for aggregate consumption to grow without any limit, even in the absence of technology. Similarly, Solow (1975) showed that a positive level of aggregate consumption can be

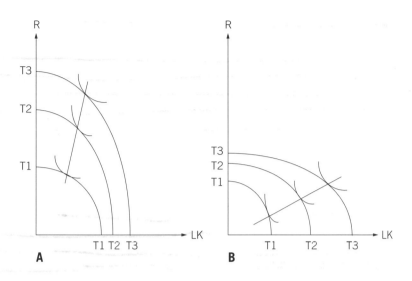

Figure 1.3   Growth models that recognise natural resource scarcity. A—outward shift of trade-off curve dominated by increased resources; B—outward shift dominated by increased labour–capital input

maintained for ever. But the main difficulty with these conclusions is that they are based on the assumption that LK can substitute for R. In reality, the substitutability between LK and environmental inputs can be very restricted. Further, environmental resources are important inputs in the formation of capital goods themselves. An added difficulty is that these substitutability theorems overlook some important implications of the laws of thermodynamics. Hence it is unwise to assume that LK can indefinitely substitute for R.

The above difficulty was at least partially addressed by Hartwick (1978). He illustrated that an economy can maintain a constant stream of consumption if it invests its returns from non-renewable (wasting) resources into reproducible capital goods that include renewable resource stocks such as fisheries and forests. This illustration has now become established as the 'Hartwick Rule'. We can simplify the Hartwick Rule in relation to figure 1.3 as follows. If an economy makes the right types of investments in renewable capital, then it will be able to maintain a given point of tangency between the output isoquant and the trade-off curve for ever.

The optimists versus the pessimists

In the review of models presented above, we observed that the classical model presents a pessimistic outcome, namely that the ability of an economy to grow has finite limits and that the economy will reach a point of stagnation. The neoclassical and the natural resource constrained models portray an optimistic outcome on the question of growth. The aftermath of the oil price shock of the early 1970s saw the revival of classical economics and the presentation of evidence suggesting clear limits to economic growth. At the same time, there was also evidence presented to the contrary, namely that economic indicators showed no sign of environmental limits. We shall now briefly consider this evidence.

The evidence for optimism

The earliest evidence for optimism appeared in the time series indexes of resource scarcity (RS) that were computed by Barnett and Morse (1963) for the period 1870 and 1957. In these indexes RS was defined as the real value (cost) of labour and capital that is required to produce one unit of real output. When measured over time, there are two distinct possibilities for changes in RS, namely that RS can either increase or decrease. An increase in RS implies the prevalence of resource scarcity. This is because we need to progressively use more labour and capital to produce the same unit of output. To draw an analogy, suppose that the output is fish. If in a given year we spend an hour of labour to catch one fish, and spend three hours of labour to do the same thing in the next year, then we can infer that fish have become scarce. Conversely, a decrease in RS implies a lack of resource (or environmental) scarcity. Barnett and Morse computed RS for specific sectors (agriculture and mining) as well as for the overall economy. Their actual evidence on RS displayed upward and downward fluctuations with a clear downward trend. Similar evidence was presented for Australia by Sinden (1972).

Evidence of this kind, namely of fluctuations in RS with a downward trend, led to the formation of the 'gap–glut' theory. The period when RS displays an upward trend is one of scarcity, and hence represents a gap. Such a gap prompts a search for technologies and innovations that will reverse the trend in RS into a downward one, resulting in a glut. Since the upward trend in RS resembles the predictions of the classical economists, the reversal of that (namely the emergence of a glut after a gap) has been referred to as a Malthusian inversion. To illustrate: soon after World War II, dairy products were in acute shortage and RS displayed an increasing trend for the agricultural sector. However, this increasing trend was soon reversed due to the introduction and spread of agricultural production technologies.

Nordhaus (1973) presented a slightly different type of evidence, which corroborated the findings of Barnett and Morse. Nordhaus estimated the real relative price of minerals with respect to labour. His observation was that the relative price of minerals had fallen over a seventy-year period (1900 to 1970). Similar observations have been made by the World Bank (1992, p. 37) by recourse to the long-run prices of non-ferrous metals between 1900 and 1991: 'Declining price trends also indicate that many non-renewables have become more, rather than less, abundant.'

If minerals are becoming scarce (due to progressively higher rates of extraction), then the relative price must rise. The fall in price is attributed to improvements in extraction technologies and methods of utilisation. For example a new car can travel a longer distance on a litre of petrol than a new car would have done, say, ten years ago. Norwegians engaged in the extraction of oil from the North Sea have developed technologies that render the extraction costs progressively cheaper, despite the fact that the oil reserve is becoming progressively scarcer.

However, some serious shortcomings in the above line of reasoning need to be noted. First, the evidence that was presented in favour of the optimistic outcomes had not accounted for the value of environmental amenities and services. For example the extraction of a tonne of coal does not merely involve labour and capital. Such extraction involves the clearing of forests, the loss of ecological systems, and the possible contamination of ground and surface water systems. Had the costs of these environmental effects been included, it is possible that the trend in RS might have been different. The second shortcoming is that these price trend analyses assumed the world market for natural resources to be competitive. This is hardly the case. Most resource producers are developing countries, and most resource buyers are the industrial countries. Michael Todaro's (2000) cogent discussion reveals that nearly ninety per cent of the resources are used up by the industrial countries. In other words, the decline in resource prices is also due to the monopsony-type purchasing behaviour of the industrial countries. The declining price trends might not have been as dramatic as perceived by the World Bank (1992) had the imperfect competition aspect been recognised. Despite these shortcomings, the evidence of the type presented above received wide acceptance, especially during the 1970s and early 1980s. One important reason for this wide acceptance was that the counter-evidence put forward by the Malthusian school was not sufficiently convincing. We shall briefly review this evidence.

The evidence for pessimism

Some of the early empirical evidence for pessimism was derived from the formulation and application of systems models that attempted to describe the reserves, linkages, and patterns of usage within a global framework. Most of the difficulties arose from the fact that these were models of the entire world. The work that received a great deal of attention was that of Forrester (1971) and of Meadows et al. (1972). Although this may run the risk of being an over-simplification, these world models can be described as consisting of three components:

- a model that describes initial conditions in terms of reserves, industrial capital stock, population size, and consumption patterns
- a system of equations describing the growth in key variables (for example resource depletion) and the linkages between key variables
- a predictor that can make use of information from the above two models to forecast the future.

These models were formulated around the late 1960s and early 1970s. The initial application was a validation of the model. This was done by setting the initial conditions at the 1900 levels and using the predictor to describe the scenario of the 1970s. This result was believed to be reasonably acceptable. The next application was to reset the initial conditions at the 1970 levels and predict the future. This prediction was a collapse of the global system—a global holocaust—due to excessive pollution, the failure of agriculture and the exhaustion of vital resources.

These dramatic findings, when widely publicised, prompted close scrutiny of the models. For example Nordhaus (1973) produced an inventory of arbitrary assumptions that dismissed the attempts of the Malthusians as 'measurement without data'. Some of the mathematical inconsistencies, for example the use of ordinary graph paper instead of semi-log graph paper to explain specific types of non-linear relationships, when highlighted, made the models and their results less credible.

Contemporary environmental issues

It was perhaps unfortunate that optimists had the upper hand in the environmental debate during the 1970s. This is because their widespread acceptance induced a certain degree of complacency in the handling of environmental matters. The major contention in developed countries was that they could fix the environment with their wealth. On the other hand, the developing countries believed that they must grow and accumulate wealth first, and that they could fix the environment later. While it is true that the rich countries are better able to deal with environmental problems than the poor countries, there are two basic flaws in the argument.

- First, even if the rich countries have a substantial amount of wealth, many environmental damages are irreversible, and there appear to be a growing number of them.
- Second, while the poverty of poor countries does not help them deal with environmental problems, there is a circular cause–effect relationship between poverty and environmental damage. Poverty prompts environmental degradation. At the same time, environmental degradation perpetuates poverty.

These observations reinforce the main focus of environmental economics, namely to achieve an equilibrium between the environment and the economy.

Australia's growing inventory of environmental problems includes: the salinisation of the Murray, the frequent algal blooms in our inland rivers, the degradation of the Darling Downs, the hole in the ozone layer, the contamination of coastal waters, and the desertification of grazing areas. The summer months in recent years have witnessed beaches in Sydney and Melbourne, and other recreational areas such as certain parts of the Hawkesbury, Murray, and Darling rivers, being closed to the public more frequently than in previous years. Almost all of this damage has been due to economic growth policies that were unaccompanied by adequate environmental safeguards. Most industrial countries share similar problems, and even items of natural heritage have been threatened. For example the Canadian maple tree and its syrup are currently under threat from acid rain. So is the famous Black Forest of Germany. Nearly half the Baltic Sea is dead owing to lack of dissolved oxygen. At home in Australia, scientists fear the long-run ecological damages that may be inflicted on the Great Barrier Reef by increased tourism and other forms of economic activity. The reef has always been an asset, and it should remain so. As the then Prime Minister Harold Holt said when the enterprising Queensland Premier Joh Bjelke-Petersen proposed drilling for oil on the reef: '... even the slightest risk to the reef is the greatest risk ...'

A similar inventory of problems has emerged in the poor countries of the world in their quest to maintain their status quo. The poor, driven into forests for their survival, have been the chief agents of deforestation. For example in the eastern Amazon basin a million hectares of forests have been cleared each year. Newcombe (1987) has estimated that the rate of deforestation in Third World countries will result in the loss of at least 4,000 biological species per year. The process of imitating the industrial countries has given the poor countries the same set of pollution problems as rich countries have. The Taj Mahal, that famous part of India's built heritage, is being gradually corroded by acid rain and atmospheric pollution. The list goes on.

The organisation of this text

The accumulation of evidence on the diverse sources of environmental damage has prompted the following thesis:

> *While environmental damage can emerge from unrestricted economic growth, it can also emerge due to poverty, unemployment, and general under-development.*

The important point that comes through from this thesis, and the evidence that supports it, is that there is a very fine balance between the environment and the economy. When this balance is lost, both the environment and the economy deteriorate. The loss of this balance can be caused by rapid, unregulated economic growth as well as by the prevalence of large-scale poverty and under-development.

This is what prompted the Brundtland Commission to decree that the environment and development are not separate challenges.

That is, economic sustainability is impossible without environmental sustainability. This is the theme with which this chapter commenced and is the theme around which the various chapters of this text are organised. We begin the next chapter with a review of how economic organisation is currently represented in the literature, namely the circular flow model of the economy. This model is the cornerstone of all economic analysis—both in microeconomics and macroeconomics. We shall consider the attempts by environmental economists to modify this description of the economic system so that the role of the environment in the economy becomes clear. We will also show that this treatment is unsatisfactory, since it does not portray the environment as the ultimate factor of production. Hence, the second chapter concludes with a hierarchical framework to describe the economic system.

Chapters 3–9 deal with the adaptation of microeconomics to the context of the revised economic system. The traditional approach in environmental economics has been to deal with environmental aspects within the frameworks of market failure, namely to internalise externalities by, for example, valuing environmental goods and resolving intergenerational issues by recourse to user costs. While retaining this framework, we shall also probe deeper into the effects of recognising the environment according to the basic tools of microeconomics, such as utility functions, indifference curves, production functions, and isoquants. For example environmental awareness can prompt consumers to distinguish between environment-friendly goods and environment-unfriendly goods, and this can affect the shape and nature of individual indifference curves. Similarly, the recognition of the environment as a factor of production could render the standard production function to be irrelevant, and the isoquants can take on a completely different shape.

Chapters 9–13 concern the adaptation of macroeconomic frameworks. We start with the ideas that underlie environmental accounting, which is an adaptation of the traditional method of national income accounting. Building on the tenets of environmental accounting, we proceed to adapt the basic frameworks of aggregate demand and aggregate supply in macroeconomics. This adaptation includes the modification of an IS–LM model and the development of an aggregate production function that recognises the environment as the ultimate resource. Such treatment will permit us to re-examine the theories of economic growth and present a theory that treats the environment and development as a common challenge.

In chapter 14 we shall return to the issue of resource scarcity that was introduced in this first chapter and examine a possible way of measuring and valuing nature at the macroeconomic level.

2

The Economic
System Revised

As indicated in the previous chapter, the goal of economic sustainability cannot be attained without environmental sustainability. Hence the economic system cannot be considered in isolation without any reference to the natural environment. If environmental sustainability must coexist with economic sustainability, then the overall system must be one that permits the identification of an equilibrium between the environment and the economy. We begin this chapter with a description of the economic system as seen in contemporary economic texts. We then consider an environmental economist's approach to adapting this description. As indicated below, this adaptation is made in terms of the effects on the economy of the laws of thermodynamics.

The standard textbook version

The multitude of economics textbooks that have appeared to date, especially the introductory to intermediate-level ones, usually include a circular flow model of the economic system. For example see Baumol and Blinder (1988), Hirschleifer (1988), and Samuelson and Nordhaus (1990). The simplest of these models explains the linkages between households and firms. This is illustrated in figure 2.1 (p. 16), and follows Hirschleifer (1988).

Households are units of consumption as well as the owners of labour and capital. The firms are units of production and supply goods and services. The households demand the goods and services produced by firms, and in turn, the firms demand the labour and capital of households. Hence two types of exchanges occur in this simple system. Firms pay wages and rents in exchange for the households'

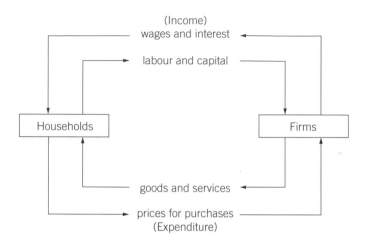

Figure 2.1   The economic system—standard textbook version

labour and capital, while households expend their income in exchange for the firms' goods and services. At the macroeconomic level, equilibrium is defined by the equality between the sum of all expenditures on final goods and services (aggregate demand) and the sum of all wages and rents (aggregate output or income). The microeconomic foundations of this model stem from the fact that each product and factor expenditure is derived from transactions in product and factor markets. More complex versions of the standard model extend the basic income–expenditure equilibrium to also include equilibria between a set of leakages and injections by recognising financial, government, and overseas sectors.

The economic system and materials balance

The environmental economists' objection to the standard description in the circular flow model is that it does not recognise the support services provided by the natural environment. Figure 2.2 illustrates a system that includes the supporting role of the natural environment. This supporting role is usually explained in terms of three functions, as follows:

1   *Provision of raw materials*: the households and the firms depend on the natural environment for air, water, and other necessities such as minerals and energy.
2   *Receptacle for wastes*: firms and households generate a large amount of wastes, which are eventually deposited in the environment.
3   *Provision of amenities*: the environment also has numerous amenities that are a source of aesthetic value. These include beautiful scenery, treks for bushwalking, and unspoilt beaches.

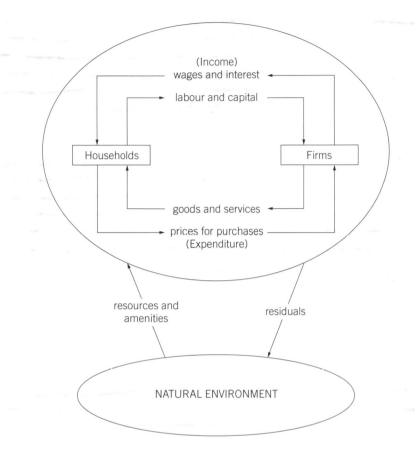

Figure 2.2　The economic system revised for materials balance

One of the earliest adaptations of the circular flow model to recognise the environment was in the development of the materials balance approach by Kneese, Ayres, and D'Arge (1970). This approach related the functions of the environment to the first law of thermodynamics, namely the law of conservation of matter and energy:

Matter can neither be created nor destroyed.

The direct implication of this law for any economy is the following equality:

| The sum of the material that enters the economy from the environment as resource flows | = | the sum of the material retained in the economy plus the sum of the residuals (wastes and pollutants) returned back into the environment | (2.1) |

At any given period, the sum of resource flows that enter the economy will not equal the sum of residuals, because the resource materials used up in the economy take time to decay before they are returned to the environment. The chief implication of this law is that residuals will accumulate in the environment and consequently prevent the environment from performing its functions. This in turn can limit and threaten the sustainability of the environment.

The limitation caused by the accumulation of residuals can be explained as follows. Suppose that an economy begins its operations in a pristine environment. The wastes and by-products of the initial set of economic activities can be assimilated by the environment, without the environment losing any of its characteristics. For example micro-organisms can decompose wastes, while plants can absorb carbon gases. This is referred to as the *assimilative ability* of the environment. But there is only a certain limit up to which the environment can display this ability. This limit is referred to as the *assimilative capacity of the environment*. When the dumping of wastes is intense and continuous, the assimilative capacity is exceeded, and the environment loses its assimilative ability and is unable to fulfil its functions as a waste receptacle. Also, when this happens, the environment, which is infested with toxic substances, also ceases to be a source of raw materials and amenities. For example in most big cities where exhaust pollution is a problem, clean air is a scarce resource and the skyline, marred by smog, is not a pleasant sight.

To summarise: the important conclusion we draw from the first law of thermodynamics is that we need to be conscious of the extent to which the environment can act as a sink for the residuals generated by the economy. If the economic system does not recognise the materials balance constraint, then profit-maximising firms will opt to discharge as great an amount of residuals as they need to. This is because the property rights of the environment are not well defined. If the residuals accumulate and the environment loses its assimilative ability, then the economic system will not be able to function.

This outcome of the first law of thermodynamics is strikingly similar to a Malthusian outcome, with the exception that Malthus predicted the limit on the grounds of a growing population. Note what Samuelson and Nordhaus (1990, pp. 854–5) had to say in the context of the Malthusian limits:

> The dour Reverend T. R. Malthus thought that population pressures would drive the economy to a point where workers were at the minimum level of subsistence … What did Malthus forget or at least underestimate? He overlooked the future contribution of investment and technology. He failed to realise how technological innovation could intervene—not to repeal the law of diminishing returns but to more than offset it. He stood at the brink of a new era and failed to anticipate that

the succeeding two centuries would show the greatest scientific and economic gains in history—a chastening fact, and one to keep in mind while listening to modern Malthusians sing on their baleful dirge.

The so-called modern Malthusians have in fact cited the accumulation of residuals, the filling up of environmental sinks and the resulting potential for losing assimilative capacity as the major factors that will limit continued growth. But do we need to be put off by their 'baleful dirge'? We can adopt the optimism of Samuelson and Nordhaus and raise the following questions:

- Can we not recycle the residuals?
- Why not find alternative uses for the residuals?
- Would not technology offer a multitude of solutions?

In other words, technology and improved knowledge can be seen as a vehicles to reverse the constraints imposed by the accumulation of residuals. However, there is another difficulty, which is imposed by the second law of thermodynamics: the entropy law. We shall consider this next.

The economic system and entropy

Entropy is the amount of energy that is *unavailable* for work. In a state of low entropy, there is substantial energy available for work, while the situation is reversed in the context of high entropy. The entropy law states that:

> Material and energy, when used in various processes, are transformed from a state of low entropy to that of high entropy.

Georgescu-Roegen (1971) and Herman Daly (1991, 1992, 1996, 1997a, 1997b) have been the main writers to relate the importance of this law to the economy. That is, the various activities of an economy draw in low-entropy resources and transform them into high-entropy residuals. Getting back resources from residuals, for example by recycling, will become progressively more difficult. Daly (1992, p. 94) describes the effect of the entropy law on the potential for economic growth as follows:

> Economies, like organisms, consist of ordered structures subject to entropic decay, but capable of building up their internal order at the expense of imposing greater disorder on their environment ... At the input end it takes low entropy matter energy from finite environment sources, and at the output end it returns high entropy wastes into environmental sinks. The sources become depleted and the sinks fill up and become polluted. The entropy law is supremely relevant because it says that sinks cannot serve as sources. Absolute scarcity arises from the combination of the first and second laws of thermodynamics, not from either alone.

Sceptics of the entropy law's relevance to the economic system have argued that this law is relevant only to a closed system, and only to energy at that. However, physicists have confirmed that the law applies to matter as well as energy. In terms of all flows excepting solar energy, the global economy and its environment can be described as a closed system. The amount of material that enters the global system from outer space is insignificant, at least at present. Even if the global system were an open one with respect to solar energy, there is a maximum upper limit to the energy it can receive during any given period, once the stocks of non-renewable energy sources have been exhausted. Hence, for all intents and purposes, our planet is virtually a closed system. This means that the potential for absolute scarcity that Daly identifies is real.

The closed nature of the materials balance system and the entropy law refutes the Dasgupta–Heal–Solow conclusions that we reviewed in chapter 1, namely that labour and capital can indefinitely substitute for non-renewable resources. It reinforces the importance of the Hartwick Rule, which stresses the need to shift towards investments in renewable capital. We need to understand, however, that shifting to renewable resources reduces the threat of absolute scarcity, but does not necessarily eliminate it. For example Brazil has adopted ethanol from agriculture as its transport fuel instead of the traditional fossil fuels. There are two things to be noted with this shift. First, intensive use of agricultural land for ethanol production can render the basic soil resource non-renewable. The other issue is that the combustion of ethanol is not without residuals, although these may be less than those generated by the combustion of fossil fuels.

The limiting influences of the thermodynamic laws do not imply that technology and new knowledge are without a role. Technology and new knowledge can definitely offset the entropic process. This can happen by substitutions between existing resources, and the development of new methods of waste treatment and recycling. For example fibre optics has replaced copper in telecommunications, and there has been a steady decline in the use of metals and energy per unit output in industrial processes. However, as Daly and Cobb (1989) argue, new knowledge is never complete. They call for caution by citing the example of asbestos which, when first discovered, was hailed as a useful resource. Years later, when the effects of asbestos fibres were understood, the usefulness of asbestos disappeared. Given our present state of knowledge, we know that the laws of thermodynamics decree that we cannot indefinitely rely on the recycling and treatment of residuals as a source of material inputs to sustain the economic system. The sustainability of the economic system also requires the economy to spend a part of its output on the environment, so that the environment is maintained. We will now consider a further adaptation of the economic system to demonstrate the economy's efforts towards safeguarding the environment.

Further adaptations of the economic system

The main difference between figure 2.1 and figure 2.2 is that the latter has included the role of the environment in terms of the material balance approach. But this description does not sufficiently deal with the effects of the entropy law. We shall now consider a further extension to figure 2.2 so that we can incorporate the thermodynamic laws and establish a basis for identifying the equilibrium between the environment and the economy. This extension is presented in figure 2.3.

The basic contention is that the natural environment is an ultimate factor of production. That is, if we try to disaggregate any good or service until no further disaggregation is possible, we will end up with a set of basic environmental factors, namely air, water, soil, and energy. These basic factors together form numerous resource systems, such as river systems, mangroves, lagoons, coastal estuaries, forests,

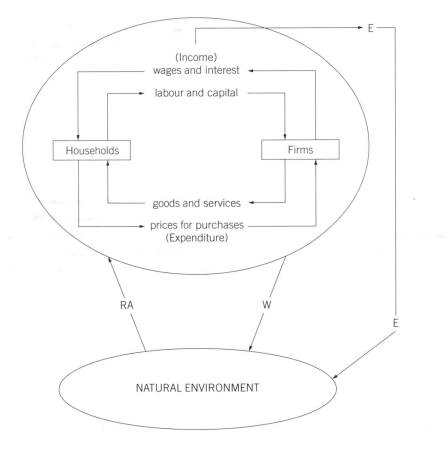

Figure 2.3 The economic system—materials balance and entropy

and so forth. These resource systems are the natural endowments that provide an economy with raw materials and amenities. They also act as sinks for the economy's residuals. In figure 2.3, the arrow RA from the resource systems to the economy represents the flow of resources and amenities; the reverse arrow W represents the flow of waste residuals. These two arrows, RA and W, demonstrate the first law of thermodynamics. If the time dimension were absent, then the difference between RA and W would represent the matter that is retained in the economy in the form of goods and services.

In the standard economic system (figure 2.1), economists assume that the sum total of all income from goods and services—final output—is used up in final consumption and investment. We can conveniently assume that the term 'final consumption' also includes government expenditure and net exports. But if environmental endowments are ultimate factors of production, then final output should not be exclusively used up in consumption and investment. A part of it must be ploughed back into the environment. In figure 2.3, arrow E illustrates the share of final output that is redirected to the environment. This share of output is for activities that will maintain, and in some instances even expand, the flow of services from the environment. The expansion of the services flow can be due to some of the activities represented by E, which strive to restore previously damaged or lost endowments and hence can expand the set of resource endowments. Other activities shown by E are analogous to regular maintenance operations—for example treating wastes before they are discharged, treating river flows at periodic intervals, and so forth. We shall consider the components of E in greater detail in later chapters, where we illustrate the adaptation of standard macroeconomic frameworks. Suffice it to say for the moment that the arrow E describes the effort expended by the economy to sustain the environment, and this effort falls within the influence of the entropy law of thermodynamics. This is because the presence of entropy results in the amount of effort entailed in E to increase as the economy's use of the natural endowments increases. Further, the effort in E will be larger when the flow in RA is dominated by non-renewables than when this flow is predominantly made up of renewables. As indicated above, technology and knowledge can dampen the rate of increase of this effort.

Equilibrium in the adapted system

We noted above that the arrow E in figure 2.3 involves two types of activities, namely:
- restoring lost (non-functional) endowments
- maintaining existing (functional) endowments.

The distinction between a functional and a non-functional endowment can be illustrated as follows. Suppose that a river cannot be used for any purpose whatsoever because it is infested with toxic algal blooms. This river is a non-functional endowment, and it cannot be counted in the stock of useful endowments owing to its toxicity. When the river is restored (by removing the algal blooms), then that restoration is like an investment. This is because the restoration adds to the stock endowments relative to what the economy started with. But if we start with a river that can be used for various purposes, then that river is a functional endowment that has to be maintained by regular clean-up operations.

The distinction between functional and non-functional endowments can be used in defining the equilibrium between the environment and the economy. In the system described in figure 2.1, economists define equilibrium as:

$$
\frac{\text{The sum of all incomes}}{\text{(national income or output)}} = \frac{\text{the sum of all final expenditures}}{\text{(aggregate demand)}} \tag{2.2}
$$

As indicated above, this definition does not explain the economy's efforts to maintain the environment. So, following figure 2.3, we can modify the above equilibrium by taking note of the differences between functional and non-functional endowments as follows.

$$
\begin{array}{l}\text{The sum of all incomes} \\ \text{(national income or output)}\end{array} = \begin{array}{l}\text{the sum of all final expenditures} \\ \textit{including} \text{ restoration of non-functional} \\ \text{endowments } \textit{less} \text{ expenditures set} \\ \text{aside to offset the wear and tear} \\ \text{of natural endowments (modified} \\ \text{aggregate demand)}\end{array} \tag{2.3}
$$

If an economy did not have any non-functional endowments—that is, if all its endowments were functional—then the equilibrium would become:

$$
\begin{array}{l}\text{The sum of all incomes} \\ \text{(national income or output)}\end{array} = \begin{array}{l}\text{the sum of all final expenditures} \\ \textit{less} \text{ expenditures set aside to} \\ \text{offset the wear and tear of} \\ \text{natural endowments (modified} \\ \text{aggregate demand)}\end{array} \tag{2.4}
$$

These modified definitions—equations (2.3) and (2.4)—describe the equilibrium between the economy and the environment so long as the efforts directed towards the environment are successful in maintaining the environment's services.

However, the feasibility of the equilibrium will depend on the magnitude of the effort and the relationship between the effort-cost and national income. For example, suppose that we are dealing with an economy where all natural endowments have been restored. The relevant equilibrium for this context is given by equation (2.4). The entropy law implies that the relationship between the 'expenditures to offset wear and tear on natural endowments', which we will denote as C_{EM}, and national income, Y, will be an increasing function. The nature of this relationship will depend on the state of the environment and the practices that have been adopted in previous years. Empirical evidence (Thampapillai & Uhlin 1994, 1997) suggests that this relationship can be an exponential function.

The relationship between C_{EM} and Y is illustrated in figure 2.4. The straight line from the origin is an 45° line that defines $Y = C_{EM}$. With the curve labelled I, equilibrium is feasible in the income range ($Y_1 < Y < Y_2$), because in this range $C_{EM} < Y$. The curve labelled II describes a highly degraded environment. Here C_{EM} is always > Y, and therefore an equilibrium is not feasible. Figure 2.4 can be also used to illustrate the relationship between entropy and technology and management practices.

In the absence of technology and environmentally sensible management practices, the curve describing C_{EM} will progressively shift to the left over time. Technology and good management practices together can prevent this shift, and can even prompt a reverse shift.

To conclude: in this chapter, we have revised the standard description of the economic system. The model in figure 2.3 describes the natural environment as the foundation stone on which the economy rests. The sustainability of the economy

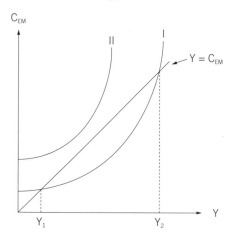

Figure 2.4   Expenditures to offset wear and tear on nature

will ultimately depend on how well this foundation is preserved and maintained. As can be seen from figure 2.3, the potential for sustainability rests on two inter-related sources, represented by arrows RA and E. If in RA we prompt a shift from non-renewable to renewable resources, then the quantity and quality of the residuals (arrow W) will become less entropic, and the amount of effort we have to apply (arrow E) to sustain the environment will also become less. By the same token, if we neglect the application of effort to sustain the environment, then there will be an accumulation of residuals and hence an increase in entropy. As a result, we will diminish the quantity and quality of resources and services that we can derive from the environment. In chapters 3 to 9 we shall examine how the associations between arrows RA, W, and E can be included in the analyses of microeconomic decisions. We shall then extend this examination to macroeconomic decisions in chapters 10 to 14.

REVIEW QUESTIONS

1 Explain why the first law of thermodynamics alone is not sufficient to explain the role of the environment in an economic system.
2 Discuss the possibility of developing a concept of equilibrium between the natural environment and the economy, especially in the context of the second law of thermodynamics.

Part 2

Microeconomics and the Environment

3

The Market Model
and Its Failure

As indicated at the end of chapter 2, in this and the five chapters that follow we illustrate how the sustainability considerations can be analysed within microeconomics. We will keep figure 2.3 as the basis for these illustrations. The main aims in this chapter are:

- to show why some or most of the components of the arrows RA, W, and E in figure 2.3 fall outside the scope of the market model
- to explore the avenues that are open to bring these components within the scope of the market model.

In other words, we will show why the market fails in the context of decisions involving the natural environment, and then sketch the ways by which this failure could be corrected.

The functions of the market

As is expounded in all economics texts, a market emerges from the interaction between the forces of demand and supply. This interaction enables the market to perform three functions, namely:

- selecting the goods to be produced
- determining the amounts of production for these selected goods
- signalling the methods of production.

These functions are more popularly expressed as: 'What to produce?', 'How much to produce?', and 'How to produce?' To explain them more clearly, we need to examine the concepts of demand and supply. In this chapter we shall confine our treatment to the bare essentials and defer the more in-depth analyses of the two market tools till chapters 5 and 6.

Market demand for a particular commodity, say apples, is the aggregate of all individual consumer demands for apples. Figure 3.1 presents a purely hypothetical society of only two consumers of apples, A and B, and shows the derivation of a market demand curve. The method of adding individual demand curves to derive the market demand curve is called horizontal aggregation. At a price of 50 cents, A consumes 2 apples and B consumes 3 apples. Hence total consumption by this society, at a price of 50 cents, is (2 + 3 = 5) apples. The market demand curve is determined by adding the individual consumptions for each price.

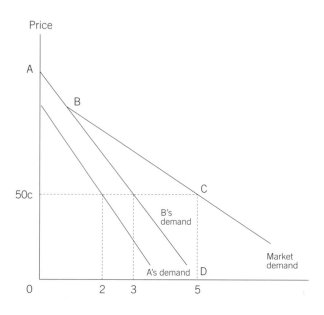

Figure 3.1   Aggregating individual demands

For each quantity purchased, an individual's demand shows the amount that he or she is willing to pay for the last additional (or marginal) unit. Since market demand is the sum of all individual demand curves in a society, it describes what society is willing to pay for a marginal unit. So for example in figure 3.1, when the hypothetical society of only two consumers (A and B) purchases 5 apples, it is willing to pay 50 cents for the fifth apple. Thus, very often the demand curve is described as a schedule of Willingness To Pay (WTP). Note that WTP is a measure of satisfaction or utility, and the WTP for the marginal unit decreases as the quantity of purchase increases. This is indeed the law of diminishing marginal utility. What is of importance to us at this point are the following:

1   WTP, being a measure of satisfaction, is also a measure of benefits; hence the market demand curve forms the basis for defining the benefits to society from consuming or purchasing a specific commodity.

2   The area below the demand curve measures the total value of WTP. For example in figure 3.1, area OABCD represents the total amount that society is willing to pay for 5 apples. So we can state that the total benefit to society from consuming 5 apples is given by area OABCD.

Consider now the market supply of apples. This is the aggregate of the individual supplies of each producer. In figure 3.2, it is assumed that there are only two apple producers, C and D, and this figure also shows how the aggregate supply curve can be derived from the individual supply curves. You will note that the method of aggregating individual supply curves to form the market supply curve is similar to the type of aggregation that was illustrated above for the demand curve. In the short run, the supply curve is in fact the ascending portion of the marginal cost function, which lies above the average variable costs.[2] Hence, the supply curve describes the marginal costs of production; that is, the cost of producing each additional unit. When prices rise, producers are willing to accept higher marginal costs and produce more. They are willing to accept lower marginal costs when prices fall. Just as the demand curve is based on the principle of WTP, the supply curve is based on the principle of Willingness To Accept (WTA). So from figure 3.2 we can infer that producers are willing to accept a cost of 50 cents for producing the eighth apple. Since the supply curve describes the marginal costs, the area below this curve defines the total cost of production. Thus, in figure 3.2, the total cost of producing 5 apples is defined by the area OEBD.

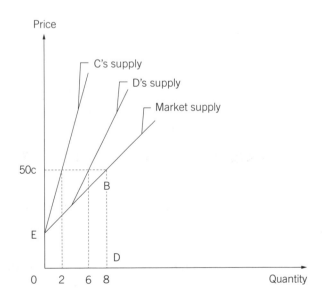

Figure 3.2   Aggregating individual supplies

The interaction between demand and supply in the market for apples in the hypothetical economy is shown in figure 3.3. The equilibrium between demand and supply indicates that Q* apples will be produced and consumed. At this equilibrium:

$$\text{Consumer's WTP for an additional unit} = \text{Producer's WTA for an additional unit} = \text{P*}$$

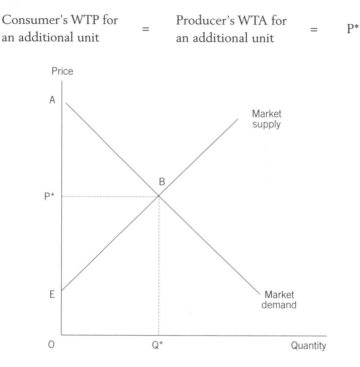

Figure 3.3 Market equilibrium

However, also note that at this equilibrium net benefits to society are at a maximum. This will become clear when the concept of net benefit is defined.

The net benefit of producing and consuming Q* apples is:

(total benefit or the total WTP for Q* apples) – (total cost of producing Q* apples)

= (area below the demand curve) – (area below the supply curve)

= (area OABD) – (area OEBD) = area AEB

If more than Q* apples were produced, we would find that the difference between the area below the demand curve and the area below the supply curve would be smaller than area AEB. This would also be the case if the number of apples produced were less than Q*.

Let us now return to the basic functions of the market mechanism, namely: what to produce, how much to produce, and how to produce. The identification of net benefits is central to the fulfilment of these functions. In a perfectly competitive economy, price is set at the point of equilibrium between demand and supply, and as we observed above, this is also the point at which net benefit is at a maximum. The market mechanism would resolve the question of what to produce by permitting the production of goods that generate positive net benefits. Clearly, if the total cost of producing any unit of a good is in excess of consumers' WTP for the good, then the market mechanism will dictate that the good had better be produced elsewhere, say overseas. The answer to the question of how much to produce is simply that the amount should maximise net benefit, and this happens when demand and supply are in equilibrium. The question of how to produce is also resolved by the size of net benefits. Suppose that a given good can be produced by two methods, where one involves higher marginal costs than the other. If demand is independent of the method of production, it is obvious that net benefits will be higher with the method that involves lower marginal costs. Hence the market mechanism signals the choice of techniques with lower marginal costs relative to those with higher marginal costs.

Environmental goods and market failure

The goods that fall within the scope of the market mechanism are called 'marketable goods' because their market demand and market supply are well defined. All marketable goods have two common characteristics:
- they can all be measured in terms of physical quantity
- they can all be valued in monetary terms.

Think of the goods at your home that are usually purchased. Bread can be measured in terms of either number of loaves or weight of a loaf in grams. Clothes can be measured in terms of the number of pieces as well as the sizes they fit. Beverages can be measured in terms of either the number of containers (bottles) or the capacity of such containers in litres. All these items also have specific prices. In a perfectly functioning market, the price of each good will equal the cost of producing the last unit of the good, and will result in an equilibrium between demand and supply. We shall now examine why this does not easily happen with all environmental goods and services.

Now consider the various types of environmental goods. Almost all environmental goods—say clean air, rivers, lakes, and forests—cannot be either measured in terms of a physical quantity or valued in money terms. Some, however, may appear to be measurable. For example one might say that a lake could be measured by its surface area or its volume of water, and that a forest could be measured by the number of trees. But note that an environmental resource cannot be considered as an entity by itself. So a lake is not simply the water it holds; rather a resource system that includes water and a range of living and non-living items. This inadequacy in terms of measurement implies that market demand and supply curves cannot be constructed, and hence a value in terms of price cannot be ascertained. The goods that present such difficulties are usually labelled 'unpriced goods' or 'non-marketable goods'. These goods present the problem of market failure, namely the inability of the market mechanism to fulfil its functions.

Several environmental goods *are* exchanged in the market: for example iron ore, coal, uranium, timber and the like. However, this does not imply that these environmental goods do not experience market failure. Environmental economists argue that these resources are often undervalued, because their valuation has invariably been based on the resource as an entity by itself and not as the component of a resource system. Mining firms price their minerals on the basis of the costs of extracting them, and these prices do not include the value of, for example, the ecology that is lost. The price of a tonne of coal is currently around $45. Does this price adequately reflect the value of trees and wildlife that have been lost or displaced in order to recover the coal? We shall return to questions of this type in the next chapter, under the headings 'Public goods and externalities', where we will show how the functions of the price mechanism are not always fulfilled, even when the market appears to work.

Price mechanism and property rights

You will find that all marketable goods will unequivocally qualify for the rules of private property rights. So first we shall consider these rules, which are as follows:

1   The right to ownership can be clearly defined. This means that it is possible to state clearly what is owned, and how much is owned. It also establishes that the good cannot be seized or used by others without the right to ownership being altered. This rule is usually referred to as the *enforceability condition*.

2   Persons other than the owner can use the good only when the owner transfers the good to others in voluntary exchange. Such an exchange also permits the ownership of the good to be transferred. This rule, as implied by its description, is called the *transferability condition*.

3   All benefits and costs from using the good are experienced only by the owner of the good. This is called the *exclusivity condition*. Others can experience these benefits and costs only after the transferability condition has been satisfied.

Now think why these rules are satisfied by only marketable goods. For the right of ownership of a good to take effect, the good itself should be clearly defined. That is, it should be possible to specify the physical quantity of the good, and that definition of quantity must encompass all aspects of the good. So a carton of milk embodies everything that is contained in that carton of milk. Now compare this with clean air or the flow of water in a river. The fact that clear boundaries cannot be drawn with many environmental goods results in violations of the enforceability condition. The transferability condition can be met only so long as there is an incentive for the transfer to take place. This incentive is price. So with many environmental goods transferabilty cannot take place (or does take place inefficiently) because the price of the good cannot be properly defined. You can also easily imagine why the exclusivity condition does not work with environmental goods. Even if you did own a lake or a river, it could be difficult for you to prevent others from deriving some benefits, at least in terms of enjoying a view. Finally, note that all three conditions are interrelated. If one of the conditions is violated, then it follows that the other conditions too will be violated.

Should the conditions of perfect competition prevail, then a system of private property rights and the price mechanism would function in unison. Such functioning results in the allocation of resources to the production of goods that maximise net benefits to society as a whole. In reality, we do not have perfect competition. Imperfect market arrangements such as monopoly, oligopoly and government intervention can hinder the maximisation of net benefits to society. Recall that in figure 3.3 we considered a society consisting of only two producers and two consumers. Although the model was overly simplistic, the market considered deals with the whole of society, and therefore area AEB in figure 3.3 can be regarded as the net benefit to the whole of society from producing apples. Also, recall that area AEB is the maximum net benefit, and this occurs in perfect competition when the producers' marginal cost is equal to the consumers' marginal WTP.

If, for example, the government intervenes and sets the price at P_L, which is well below P^*, then consumers will demand Q_L, but producers will supply only q_L, as shown in figure 3.4. Hence a shortage of $(Q_L - q_L)$ will occur in the market. Shortages often drive prices up, due to hoarding by sellers, and the price could settle at P_u when q_L is made available in the market. In this context, the total WTP for q_L units is the area below the demand curve at q_L, namely area $OADq_L$. The total cost of producing q_L is the appropriate area below the supply curve; that is, area $OECq_L$. The resulting net benefit (total WTP – total cost) is area ADCE, and this is clearly smaller than the equilibrium net benefit of area ABE.

The shortcoming caused by imperfect competition is further aggravated by the inability of the price mechanism and the system of private property rights to cope with environmental goods—that is, market failure. The aim of national policy is to choose strategies that will enhance overall net social benefits. Hence the strategies that governments could consider fall into two categories: those that reduce the extent of imperfections to make the markets more competitive, and those that attempt to remedy the effects of failed markets.

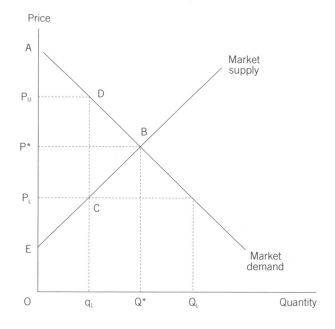

Figure 3.4   Imperfect competition

In general, there are two broad approaches to correcting the problem of market failure that occurs with environmental goods.

- The first is to somehow measure and value environmental goods so that a market for them can be created.
- The second is to allocate the private property rights of environmental goods to persons either individually or collectively, so that a market will emerge from such an allocation.

With respect to the second approach, the argument is simply as follows. When the ownership of a capital asset is vested with an individual, that person will adopt strategies to ensure that the value of the asset is maximised. With almost all items of capital, the maximisation of asset value also coincides with the maximisation of the life of the asset. Environmental goods can be regarded as items of capital. When their ownership is vested with private individuals, then those owners will adopt strategies to maximise the value of environmental capital—that is, to sustain the environment. This reasoning is the basis for the property rights (privatisation) approach to restoring market failure. Proponents of this approach argue that the Black Forest in Germany has survived centuries of use primarily because the forest was collectively owned by a group of families with well-defined rules of ownership and utilisation. Blyth and Kirby (1985) illustrate the positive role of private ownership with reference to Australia, where a significant proportion of farmland is under state ownership on leaseholds. They indicate that farmers who own their land outright or those who have no uncertainty regarding the renewal of their leases carry out soil conservation works to a much higher standard than those whose ownership and prospects of lease renewal are uncertain. However, in practice the privatisation approach has several difficulties, which are encapsulated in questions such as: Who should get the property rights? If they go to the highest bidder, will not the prevailing inequalities of society be aggravated? For this reason many governments have been prompted to adopt commercial approaches to the management of environmental goods. For example see box 3.1.

Regardless of which of the approaches is adopted (that is, restoring the market or allocating property rights), the valuation of the environmental goods is central to both. For example, with the restoration of markets, the valuation of environmental goods permits us to define the demand for, and supply of, environmental goods. With the allocation of private property rights, the valuation of environmental goods permits recognition of asset values. The development of methods for valuing environmental goods becomes relatively easy when these goods can be placed into specific conceptual categories, namely as public goods and externalities. We shall consider these next.

Box 3.1 Property rights for environmental goods

Although in theory the privatisation of environmental goods may sound attractive, in practice it is difficult to get the private sector to take charge of these goods. Therefore, in many instances governments have set up corporations or similar entities to commercially manage them. A good example is Laboe, which is a coastal suburb in the port city of Kiel in northern Germany. Located at the mouth of the Kiel fjord, which opens into the Baltic Sea, Laboe is a popular tourist destination, not only for the Germans but also for many Europeans. Certain beaches have been fenced off and the local government operates these beaches just like a private business. To gain access to these beaches and their facilities people are charged an entry fee. The commercialisation of the beaches has proved successful in two ways. First, the levy of entry fees has reduced congestion on beaches that would have otherwise suffered from excessive usage, and the income from the entry fees has been used to maintain the beaches with a high level of care. Second, this high level of maintenance and the growing popularity of beaches have prompted the port authorities to tighten controls on the ballast discharges of ships that use the Kiel harbour. In Australia, Waste Services NSW is a government corporation that has taken over the responsibility of managing several of our environmental resources (river systems, oceans, and land-fills) that serve as receptacles for liquid and solid wastes. The commercialisation of the management of environmental resources for waste disposal has facilitated the emergence of recycling and site remediation as ancillary commercial activities.

The commercialisation of environmental goods is increasingly being viewed in a global sense as well. In May 1999 Duke University hosted the Fourth Annual Cummings Colloquium on Environmental Law. The theme of the colloquium was: 'Global Markets for Global Commons: Will Property Rights Protect the Planet?' Some of the main questions explored in this colloquium were: can new global property laws and market-based institutions be successfully designed and implemented to constrain overuse of the global environment? Can global environmental resources be 'propertised' in effective ways, so that global environmental conservation and stewardship is 'internalised' into global markets and can be 'purchased' by its beneficiaries? Can property rights in the global environment be created under international treaty law? Could traditional property rights—even to goods now traded in global markets—have evolved locally? What institutions are needed to create and supervise new global environmental property markets, ensuring both efficiency and fairness?

REVIEW QUESTIONS

1 Explain how and when the market and a system of private property rights can work together to maximise net social benefits.
2 Box 3.1 contains examples of beaches and waste receptacles that are commercially managed. Try illustrating the possible shapes of supply and demand for these environmental goods.
3 Discuss the following statement: 'The privatisation of environmental goods is an important means of achieving sustainability.'

4

Public Goods
and Externalities

As indicated in the previous chapter, the valuation of environmental goods and services is central to correcting the problems of market failure. In this chapter we consider the conceptual frameworks for two types of environmental goods and effects in order to render this task of valuation somewhat easy. Recall figure 2.3 in chapter 2 and the two sets of arrows RA and W linking the environment with the economy. Most environmental goods that can be categorised as public goods tend to belong to the arrow RA. That is, they can be regarded as service flows from the environment to the economy. Most of the environmental effects that are externalities are generally included in the arrow labelled W. These will be considered next.

Public goods

A pure public good is one where a person's consumption of the good is not rivalled by any other person's consumption of the same good. This non-rivalry in consumption means that if a person consumes a public good and derives benefits from it, there is no reason to suppose that the benefits to other consumers of consuming this public good will be reduced. Pure public goods are indeed rare. The usual textbook examples are a lighthouse and defence. Suppose that several ships are approaching Sydney harbour. The benefits of direction and protection that are afforded by the signals from the lighthouse are not diminished for a ship because the other ships have also seen the signals. Defence is similar. Suppose that the scientists working for the Commonwealth Department of Defence have constructed a super-missile that can strike and destroy any enemy missile, vessel, or aircraft that approaches Australia's sovereign territory. (Please note that this situation is purely imaginary—Australia does not have any enemies and is a peace-loving country.) Once this missile has been constructed, each person living in Australia will feel protected without anyone else's individual sense of protection being diminished.

From the above two examples, we see that public goods have three important properties. These are as follows:

1   *Non-rivalry in consumption.* The consumption of the good by one person does not diminish the benefits of the good enjoyed by another person.
2   *Non-diminishability in consumption.* One person's consumption of the good does not diminish the total availability of the good.
3   *Zero marginal costs.* This property follows from the first two. If we cannot deny the benefits of the good to anyone and also cannot reduce the quantity of the good, then the cost of providing the good to an extra consumer is zero.

Note the distinction between private goods and public goods. With private goods, we have rivalry in consumption and positive marginal costs for additional users. This textbook is a private good. Supposing that there is only one book left on the shelves of the bookstore, and you purchase this last copy, then another reader will be unable to derive satisfaction from the book. So with private goods the availability of the good diminishes with consumption. If the publishers wish to make extra copies of this book, then they have to bear positive marginal costs.

Many public goods also display the characteristic of non-excludability. It is not possible to prevent someone from consuming the public good, even if that someone is not willing to pay for it. This is clearly the case with a public good such as national defence. Katz and Rosen (1991) argue that excludability and non-excludability are primarily a result of technology and legal arrangements and are not due to the public nature of a good.[3] For example, suppose that a defence shield protects a specified but large land area of Australia. Should the residents of Australia perceive imminent danger, then it is possible for everyone to flock into this shielded area and feel protected. However, it is also possible for a draconian law to prohibit some inhabitants from entering the protected area. That is, despite the publicness of the good, exclusion is possible through a legal arrangement. But it is of course true that many goods that display non-rivalry in consumption also display the property of non-excludability.

Most environmental goods are 'quasi-public goods' or 'mixed goods'. This means that they display public good properties up to a certain point, beyond which they become private goods. Consider a national park. A relatively large number of visitors can enjoy the amenities of a national park without diminishing the enjoyment of one another's use of these amenities. But this is true only so far. As illustrated in figure 4.1, the marginal cost of providing the amenities of a national park is zero for up to q_1 visitors. But when the number of visitors exceeds q_1, the marginal cost begins to become positive and tends to infinity when the national park becomes congested.

We can offer a similar explanation for the public good property of a range of other environmental goods such as lakes, rivers, and clean air.

It may now be reasonably clear how the public good property creates market failure. First, public goods present a pricing problem. The market solution is zero price. This is because the supply (marginal cost) curve is along the horizontal axis, at least for the most part, as is shown in figure 4.1. With public goods that are gifts of nature, the zero price condition prompts excessive use, resulting in depletion and deterioration. For example, for a long time industrial pollution was freely emitted into the atmosphere and waterways, because these were regarded as free receptacles. Up to a point, the natural receptacles of air and water are capable of assimilating the pollution. For example micro-organisms can degrade some pollutants and plants can absorb excess carbon dioxide. However, just as congestion can exclude people from enjoying the amenities of a national park, excessive pollution can prevent, and in many instances certainly has prevented, further use of the natural environment as a receptacle for pollution.

The second cause of market failure with public goods is also associated with zero price. It is difficult to demonstrate a true demand curve for a public good. This is because people will tend to understate their WTP for a public good. This in turn is because people realise that, once it is provided, they can have it free. This problem is often referred to as the 'free rider' problem.

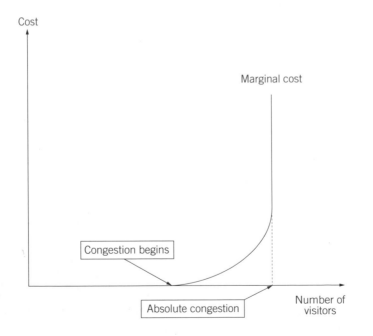

Figure 4.1   Quasi-public good characteristics of a national park

As indicated previously, we can consider one of two approaches to correcting the market failure caused by public goods. The property rights approach has been common with infrastructure public goods such as tollways and bridges, and has been extended to some environmental public goods such as forests, beaches, and national parks. The usual property rights approach is to set a reservation price and then auction the rights to manage the public good with some pre-specified rules of management. The highest bidder who acknowledges the pre-specified rules of management and exceeds the reservation price wins the opportunity to manage the public good. However, the valuation of the public good is still inevitable because a reservation price has to be set.

Externalities

An externality creates an interdependence between two or more groups of people, and this interdependence is unpriced. Consider a factory that produces steel and also emits smoke into the atmosphere. Both steel and smoke create interdependence between the factory and those who live in the areas surrounding the factory. But steel is not an externality, because if the people living around the factory wished to use the steel, they would have to make a monetary payment to get it. That is, the factory would be compensated for producing a useful product. On the other hand, the type of interdependence that is caused by smoke is one of disutility, namely discomfort, illness, and so on, and cannot be readily compensated. If we decide that the factory must compensate its nearby residents for the emission of harmful gases, then how do we decide on the size of the compensation? This is not the case with steel, where the size of the compensation can be ascertained by estimating the various costs of production, and is in fact the market price of steel. So we define smoke as an externality because, apart from creating an interdependence, it is also unpriced or uncompensated. We shall return to the issue of compensation subsequently, when we look at the various policy options. However, for the moment we shall consider some examples and illustrate how externalities cause market failure.

The third runway at Sydney's Mascot airport has been a source of controversy for a long time. The additional runway, which was completed in 1994, represented significant cost savings in that the need for building another airport elsewhere was substantially reduced. This also eased a large amount of air traffic congestion and has also perhaps permitted an increase in the number of scheduled flights in and out of Sydney. An increase in air traffic would make Sydney a popular airport for travellers and people in business, and would generate additional government revenue (due to the collection of more landing taxes and so on). However, the third runway has also caused significant disutility to the residents of the suburbs that surround the airport. This disutility is specifically due to the increases in the noise level, which in turn are due to a larger number of aircraft operating out of Sydney than before. So noise pollution has created interdependence between those who find the additional runway convenient (for example people in business) and those who live near the airport. Since this interdependence cannot be easily compensated, noise pollution is an externality.

Another externality that is causing much concern in Australia is the salinity of urban water supplies due to agricultural practices. The classic example is the externality caused by intensive irrigation in the Murray–Darling Basin. Prolonged intense irrigation has resulted in the raising of the water table in the basin, and this in turn has resulted in the increase of salt intake from groundwater sources. The adoption of agricultural drainage practices to alleviate the problem of soil salinity has resulted in the discharge of salts and agricultural chemicals into the river system. This has in turn caused a deterioration in the quality of water supplies to the city of Adelaide some thousand kilometres downstream. In this example the externality is the contamination of urban water supplies, and this externality creates interdependence between the inhabitants of Adelaide and the farmers of the Murray–Darling Basin.

Note that, in the examples given above, the term 'interdependence' was used, though the term 'conflict' might have been more appropriate. This is because not all externalities create disutility. For example, if one or two people along a street maintain a beautiful garden, it could prompt the others along the street to follow suit. As a result the whole street receives a facelift. Take another example. Farmer Jones may build a dam upstream to protect his property from flood damage, but the benefits of flood protection will also spill over to farmers John and Robert, who live downstream.

Consider now how externalities cause market failure. In figure 4.2, we have hypothetical demand and supply curves for agricultural output from the Murray–Darling Basin. The usual supply (marginal cost) curve does not account for the externality of water pollution. Suppose that we have been somehow able to estimate the monetary value of the disutility caused by water pollution. This is shown as the *marginal externality cost* (MEC). If we add the marginal externality costs to the marginal costs of producing agricultural output, then we will have a supply curve that includes the externality. The supply curve that incorporates the externality is usually referred to as the *marginal social cost* (MSC) because it includes the costs incurred by all parties, who in this case are the agricultural producers and the urban dwellers. The curve that omits the effect of the externality is called the *marginal private cost* (MPC), as it deals only with the costs incurred by one group, namely the agricultural producers.

Now note that the market solution based on the MPC results in excess agricultural output, while that based on the MSC lowers output and raises the price.

So the market solution (P$_2$, Q$_2$,) that is based on the MSC is often referred to as a *social optimum*. However, the attainment of a social optimum is possible only if the externality can be valued in monetary terms and included in the demand and/or supply curves. In this example we illustrated the inclusion of the effect of the externality in the supply curve only. It is also possible that consumers may lower their WTP for the agricultural output due to their awareness that agricultural production is contributing to water pollution. The inclusion of the effects of the externality in the components of the price mechanism is termed the internalisation of the externality. However, given that the externality is an unpriced item, the feasibility as well as the effectiveness of internalisation would depend on the types of methods that can be developed to value unpriced effects.

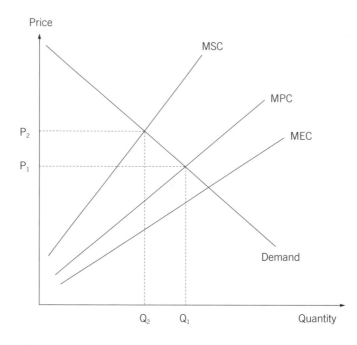

Figure 4.2   Illustration of an externality

The relationship between public goods and externalities

To summarise, we commenced the previous chapter with a description of the price mechanism. We illustrated that, under conditions of perfect competition, the price mechanism and a system of private property rights operate in unison, resulting in a market solution that maximises net benefits to society as a whole. However, with environmental goods, the price mechanism and the system of private property rights both fail. This is because environmental goods cannot be easily measured and valued. This unpriced characteristic of environmental goods can also be explained in terms of two properties they possess, namely the properties of public goods, and the generation of externalities whenever they are utilised.

Hence with several environmental goods it is possible to illustrate a relationship between externalities and public goods. It is very often the interdependence between persons that represents the externality that causes the degradation of a public good. For example as consumers we use detergents to wash, clean, and brighten our clothes. The phosphorus in these detergents finds its way into our water supply, in rivers and dams. Phosphorus is also found in the residual material that is discharged into waterways from sewage treatment plants. Phosphorus, together with nitrogen, is the major cause of the proliferation of Cyanobacteria, which manifest as toxic blue-green sludge. This blue-green sludge is the externality that causes interdependence between the members of a community, because the community has used a public good (waterways) to discharge its residues. Note that the externality occurs because the public good has become degraded.

We illustrate this in figure 4.3. Panel A of figure 4.3 illustrates the standard description of a quasi-public good as is found in most microeconomics texts. Suppose that in this instance, the public good is the air-shed of a region where certain polluting industries are to be located. These industries would use the air-shed by emitting smoke into it; that is, the air-shed would be used as a sink for the various emissions. As the number of users of the air-shed—namely the polluting industries—increases, the quantity of emissions also increases. This is shown in panel B. In panel B, the emission quantity of W_{AC} represents the assimilative capacity of the air-shed. That is, the air-shed is capable of tolerating emissions up to the quantity of W_{AC} without losing any of its properties, such as the proper composition of atmospheric gases. Consider panel A. The marginal cost of letting an extra user emit pollution is zero, until the quantity of emissions reaches the assimilative capacity of the upper atmosphere. This corresponds to the number of users being equal to N_{AC}. As the number of users increases, the amount of pollution also increases and at some critical quantity of emission, W_C, which is caused by N_C users, the air-shed becomes fully saturated and ceases to be a sink. At this point (N_C) the marginal cost of an extra user in panel A tends to infinity.

Now consider panels C and D. In panel C, the relationship between the quantity of industrial production and the quantity of pollution is presented, and panel D describes the standard market for the industrial product. Note that, in panel D, the non-recognition of the externality cost (C_E) leads to a market solution of (Q_P, P_P) and is defined by the equilibrium between demand and the industry marginal cost (MC). The social marginal cost (SMC) is the sum of the MC and the externality cost, which emerges when the quantity of production exceeds Q_{AC} due to waste production exceeding the level of assimilative capacity. As a result, the SMC deviates from the MC when the quantity of production exceeds Q_{AC}.

The externality cost and SMC both tend to infinity at production quantity Q_C, which corresponds to a pollution quantity of W_C and user quantity of N_C.

The standard argument for government intervention stems from the fear that, if unregulated, the externality cost will be ignored and production will settle at Q_P. This is not too far from Q_C. Further, with increased demand and a mixture of increased production and new entrants, production quantity can in fact reach Q_C, at which level there can be an environmental disaster. Governments would like to reach Q_S by internalising the externality. This involves valuing the externality and being able to quantitatively specify the externality cost function. Should this be

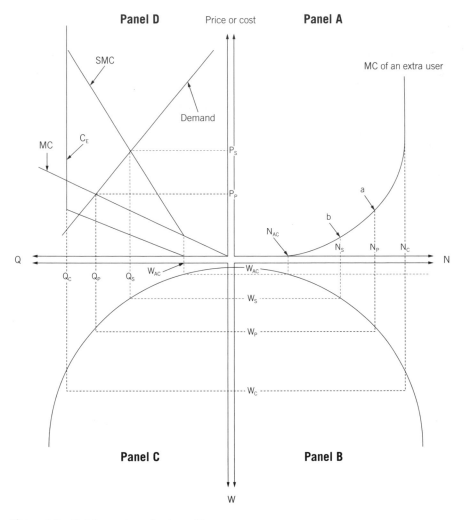

Figure 4.3   Public goods and externalities

the case, governments would impose a tax that is based on the price differential
($P_S - P_P$) or impose a quota restriction on the product market to Q_S. The effect of
this on the demand for the public good is readily apparent. When unregulated,
the demand for the public good is at point **a** in panel A. When the government
intervenes, this demand is moved to point **b**, which corresponds to quantity Q_S in
panel D. As this conceptual example illustrates, the need for government intervention and regulation stems from the need to protect the air-shed and maintain
the flow of services it offers. Note that this flow of services is important not only
to the industries that use the air-shed, but also to the wider public.

To generalise from the above illustration, an important policy directive to safeguard environmental goods is to regulate the use of the public good; that is, to
institute controls on the discharges into waste receptacles such as waterways and
air-sheds. As we shall see later, such controls may be achieved through taxes, standards, or a combination of both. In some instances it is also possible to have control
by measuring the quality of the commodity that is responsible for the externality.
This too could be achieved through standards and penalties. It is this type of control that has prompted the emergence of 'phosphate free detergents' and a range
of other consumption goods with environment-friendly attributes.

As demonstrated in this chapter, the valuation of environmental goods is essential for bringing these goods into the market framework. The logical step then is to
consider how environmental goods can be valued in monetary terms. However,
before doing this, it will be useful to examine some underlying theories of consumer and producer decision-making, and how these theories should be adapted
to the context of environmental goods. An understanding of these theories can also
help in the development of methods for valuation.

REVIEW QUESTIONS

1 Consider figure 2.3 in chapter 2. Why do most of the items that could be
   included in R and A border on being classified as public goods and those that
   could be included in W be regarded as externalities?
2 How do public goods and externalities cause market failure?
3 Provide an example that illustrates the relationship between a public good
   and an externality. Use an appropriate supply–demand framework to show
   how the overuse of the public good becomes an externality.

5

Consumer Demand and the Environment

In this chapter we return to consumer demand, which is a basic tool of the market model. We shall explore this tool more closely because the relationships between environmental issues and consumer demand have become important in policy formulation at both the social planning level and the individual business level. For example, as we saw in the previous chapter, the specification of environment-friendly attributes for consumer goods is an important role of regulatory bodies such as Environment Australia and the Environment Protection Agency. Business managers are also actively seeking ways to design goods that display these environment-friendly attributes because this helps them capture a larger share of the market.

We shall first summarise the salient features of consumer demand theory as they are exposited in most standard texts. We shall assume that the reader has a sufficient grasp of these topics not to require in-depth treatment of them here. They are:

- the underlying conceptual premises of consumer demand, namely utility functions, indifference curves, budgets, and substitution effects
- factors influencing a demand function such as prices, income, and advertising
- elasticity of demand and business profits.

In order to explain the linkages between consumer demand and the environment, three considerations are introduced here.

- The first is the illustration of how increased consumer awareness of environmental issues could be explained through the theory of indifference curves in conjunction with some observations from behavioural economics. This involves a concept called the endowment effect and the disparity between willingness to pay and willingness to accept.

- It is possible to show that the market demand curve, which most decision-makers study for the purposes of developing their strategies (for example product development by business managers and regulation by social planners), can be easily influenced by the environmental awareness of consumers as well as the environmental attributes of the good in question.
- In just the same way as information on price and income elasticity of demand is important for both social and business decision-making, it is pertinent to argue that information about some new concept such as *the consumer elasticity of environment-friendliness* could prove useful when decisions involve environmental considerations.

All these considerations relate to the topics that we are about to examine. We shall deal with these in turn.

Utility functions, indifference curves, and demand

The standard theory

In standard microeconomic theory one assumes the existence of a rational, utility-maximising consumer who is constrained by a fixed budget. When the underlying utility function of a consumer is one that displays diminishing marginal utility, and the consumer's consumption basket is limited to two goods, it is possible to demonstrate a convex set of indifference curves, as shown in figure 5.1 (p. 50). As is explained in most standard texts, the consumer maximises his or her utility at the point where the budget line is tangent to the highest attainable indifference curve. Further, standard theory also illustrates how a price consumption curve can be derived for one of the two goods. For example consider good B in figure 5.1. By holding the price of good A constant and varying the price of good B, it is possible to trace a locus of tangencies between the budget line and the highest feasible indifference curve. This locus provides the basis for the demand curve.

In common language, the demand curve explains an individual's willingness to pay (WTP) or willingness to accept (WTA) for an extra unit of a good. This WTP or WTA decreases as the quantity of consumption increases, due to the law of diminishing marginal utility. The theory of indifference curves explains this relationship quite elegantly.

Although decision-makers are ultimately interested in the nature of the demand curve and its responsiveness to pertinent parameters, the underlying utility function and the indifference curves themselves provide useful information for decision-making. For example the slope of the indifference curve measures the marginal rate of substitution (MRS), and information concerning MRS can be useful for product

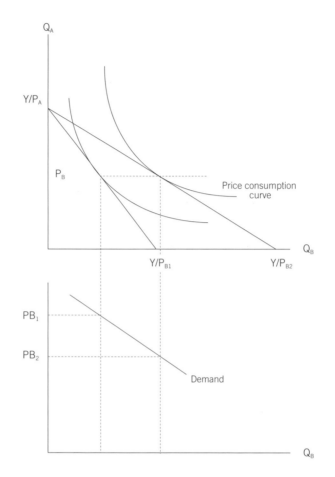

Figure 5.1   The derivation of the demand curve

development. However, from an empirical point of view, demand equations are derived directly from market and household surveys using econometric methods and decision-makers gain their information on substitution behaviour through the concept of cross-elasticity of demand, which we will consider later.

The reason we consider the theory of indifference curves here is because this theory needs some adaptation to account for certain findings within the subdiscipline of behavioural economics. What is relevant to us here is the finding that *an individual's WTP for a good is consistently less than his or her WTA for the good. This disparity between WTP and WTA appears to be more marked when one considers substitution behaviour between an environment-friendly good and an environment-unfriendly good.* At the centre of this disparity is a consumer attribute that some economists have labelled the endowment effect. We shall look at this next.

Box 5.1 Environment-friendly attributes of consumer goods

Have you checked the labels of goods on your supermarket shelves? Many producers attempt to capture consumers' attention by highlighting the environment-friendly attributes of their goods. Some of these are:

- dolphin-safe tuna
- teabags made from unbleached paper
- shoes from recycled raw material
- biodegradable detergents
- organic farm produce
- recycled paper.

  And the demonstration of environment-friendly attributes is not confined to the supermarket shelves.

- DaimlerChrysler, the European automobile manufacturer, claims that its manufacturing process reduced water consumption from 35 cubic metres per vehicle in 1992 to almost 5 cubic metres per vehicle in 1998.
- Many Australian firms have entered into 'Eco-Efficiency Agreements' with Environment Australia (EA), the Commonwealth agency that oversees environmental management and policy. These are voluntary agreements between EA and industry associations in the pursuit of ways to minimise adverse environmental impacts of production.
- Many electricity retailers (including those in Australia) now offer green tariffs to consumers in order to increase the renewable energy content of electricity. The offer of these green tariffs and the capability to generate green power is seen as a means of attracting environment-friendly consumers.

Adapting the standard theory for environmental effects

To date, the disparity between WTP and WTA has been reported by several authors (Bateman et al. 1997; Franciosi et al. 1996; Kahneman, Knetsch, & Thaler 1990; Knetsch 1989, 1995; and Knetsch & Sinden 1984). In the literature, the cause of the disparity has been largely attributed to the *endowment effect* (Kahneman, Knetsch, & Thaler 1990; and Knetsch 1989). The term 'endowment effect' refers to consumption behaviour stemming from an individual's possession of a good. Generally, the effect is described as one where individuals are less willing to part with a good after taking possession of that good relative to their willingness to part with it before they acquired it. With a set of simple assumptions it is possible to show how the endowment effect can be internalised into the theory of indifference curves. Then the derivation of demand curves within the framework of substitution behaviour, utility maximisation, and a budget constraint will show the disparity between WTP and WTA (Thampapillai 2000). Important for us of course is the existence of this disparity when a consumer is substituting between an environment-friendly (EF) good and an environment-unfriendly (EUF) good. The discussion below is taken from Thampapillai (2000).

Pertinent assumptions

Suppose that a consumption basket can be regarded as an endowment, and consider varying endowments involving two goods A and B where A is an EF good and B is an EUF good. For example a person could have become accustomed to consuming 4 units of A and 6 units of B on a regular basis; this consumption pattern (basket) is regarded as similar to an endowment. As in standard theory, an individual displays a specific set of indifference curves, but the shape of these curves is governed by the individual's endowments. Specifically, the shape of the indifference curves will be such that the individual places a higher value on the loss of a good than on the gain of the other with reference to the endowment. To illustrate, consider figure 5.2, in which three different endowment positions **a**, **b**, and **c** are shown. For consumption behaviour involving the endowment at **a**, an individual displays a set of indifference curves labelled $(I_1, I_2, \dots)$ as opposed to the set of indifference curves labelled $(II_1, II_2, \dots)$, which describe consumption behaviour for the endowment at **b**. That is, the shape of the indifference curves $(I_1, I_2, \dots)$ is such that the loss of A commands a higher value than the gain of B,

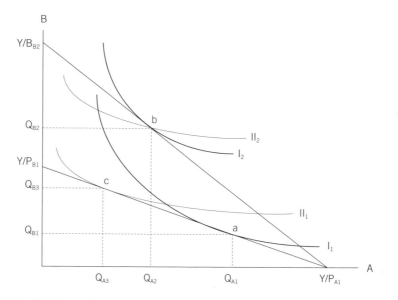

Figure 5.2 Varying endowments and indifference curves

and the shape of indifference curves $(II_1, II_2, ...)$ is such that the loss of B commands a higher value than the gain of A. This distinction in valuation can be explained by differences in the marginal rate of substitution (MRS) and the ratio of marginal utilities (MU). For example, for the indifference curves $(I_1, I_2, ...)$:

$$\{(MRS_{AB}) = (MU_A)/(MU_B)\} > \{(MRS_{BA}) = (MU_B)/(MU_A)\} \qquad (5.1)$$

The above inequality is reversed for indifference curves $(II_1, II_2, ...)$. To generalise: for two goods A and B, should an individual's endowment contain more of A than of B, then the individual's underlying utility function is such that $\{(MRS_{AB}) > \{(MRS_{BA})\}$.

The contention here is that the shape of an individual's underlying utility function, and hence the indifference curves, changes when the individual's endowment changes. That is, should an individual be prompted to shift from a to b in figure 5.2, then the individual's indifference curves will also change from $(I_1, I_2, ...)$ to $(II_1, II_2, ...)$.

The elicitation of demand curves

Consider an individual who allocates a fixed budget of $Y towards the consumption of two goods A and B, the prices of which are respectively P_{A1} and P_{B1}. Assume that utility maximisation subject to the budget constraint occurs at point **a** in figure 5.2: that is, the point of tangency of the indifference curve I_1 with the budget line that satisfies $\{(P_{A1}{}^*A) + (P_{B1}{}^*B) = Y\}$. The consumption basket (or endowment) at this point is $\{(A = Q_{A1}), (B = Q_{B1})\}$. Suppose now that the price of B falls to P_{B2}, resulting in a new budget line that satisfies $\{(P_{A1}{}^*A) + (P_{B2}{}^*B) = Y\}$. The utility-maximising consumer now changes his or her endowment to point **b**. The new endowment is defined by $\{(A = Q_{A2}), (B = Q_{B2})\}$.

The new endowment $\{(A = Q_{A2}), (B = Q_{B2})\}$ causes the consumer to form a new set of indifference curves, namely $(II_1, II_2, ...)$. Should the price of B now revert back to P_{B1}, the consumer would alter his or her consumption behaviour to maximise utility at point **c**, which is the point of tangency of the indifference curve II_1 with the original budget line, namely the one that satisfies $\{(P_{A1}{}^*A) + (P_{B1}{}^*B) = Y\}$. The revised endowment in figure 5.2 is $\{(A = Q_{A3}) (B = Q_{B3})\}$. This new endowment will result in the formation of a new set of indifference curves, and the process of changes in the utility function due to changing endowments will continue in this way.

Consider the consumer's movement from **a** to **b** to **c**. The endowment effect results in the demonstration of two price consumption paths, namely a–b for the price change of B from P_{B1} to P_{B2}; and b–c for the price change of B from P_{B2} to P_{B1}. The price consumption path a–b depicts a demand curve based on the WTP for B, because the path is based on the individual's acquisition of B.

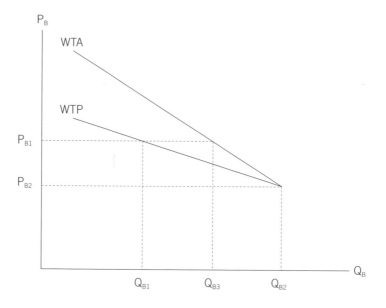

Figure 5.3   Two distinct demand curves

The path **b–c** depicts the demand curve based on WTA for B because it is based on the individual's sacrifice of B. These are illustrated in figure 5.3.

It is possible to conceptualise that the distinction between the two demand curves (and between the shapes of the underlying indifference curves) would diminish if the two goods A and B became increasingly similar, and vice versa. So, for example, if a business manager is contemplating capturing a larger share of the market through product development, it is useful for the manager to consider:

- product development strategies that would clearly distinguish the good from those offered by rival firms
- marketing and advertising strategies that would clearly inform the consumer of the distinguishing attributes of the good.

In the context of a society where environmental awareness is on the rise, pursuing product development involving superior environmental attributes and marketing them to inform consumers will no doubt be a sensible business decision. It may also be pertinent for businesses to initially lower the product price of an EF good relative to the price of the rival EUF good. This is because the sequence of events that follow a price reduction for an EF good (as shown in figure 5.2) indicates that a consumer will eventually settle on consuming more of the EF good, even after the original price is restored. Similar reasoning can be given to justify government intervention with price incentives to encourage the consumption of EF goods.

The market demand curve and environmental effects

From the above discussion we can see that it is possible for a decision-maker to carry out WTP and WTA surveys to infer the merits of pursuing a particular strategy (such as product development or portfolio diversification). We can now show the relationship between the above discussion and the standard demand equation, which is often estimated from market surveys through econometric methods. These techniques are often not capable of capturing the distinction between WTP and WTA, because the data for the analysis are often taken from actual recorded market transactions.

For convenience, suppose that the demand equation is linear and is as follows. Note that there can be several more variables that would influence demand than are considered below. However, for purely illustrative purposes, we limit these to six variables that are generally considered in most economics texts.

$$Q_D = \beta_1 P + \beta_2 P_A + \beta_3 I + \beta_4 A + \beta_5 T + \beta_6 N, \qquad (5.2)$$

where

Q_D represents the quantity demanded
P is the price of the good
P_A is the price of the alternative good
I is income
A is advertising
T stands for tastes or attitudes
N is population
the β coefficients measure the responsiveness to each variable.

Most standard texts will show how changes in P explain the movement along a given demand curve, while the remaining variables in equation (5.2) explain shifts in the demand curve. For example if the price of the alternative good falls, then the demand for the good in question is likely to shrink, or if consumers' disposable incomes rise, then the demand for this good will increase.

If, for example, a business manager or policy-maker is exploring the potential for promoting a good with favourable environmental attributes, then in the first instance the survey data on tastes and attitudes could focus on the consumers' environmental awareness. The demand curve for an environment-friendly good is most likely to shift to the right as consumers become more aware of its environment-friendly attributes. Further, if the business manager expects that the environment-friendly good has to be sold at a higher price, then he or she will want to learn more about the possible changes in incomes and attitudes of the clientele. This is because increases in income are often associated with higher demand for better quality and improved attitudes towards the environment.

Elasticity of demand and the environment

Elasticity of demand measures the sensitivity of quantity demanded to specific variables such as price and income. It is defined as the *percentage change in quantity demanded in response to a one per cent change in the selected variable.*

Price elasticity of demand explains the sensitivity of demand for the good with respect to the good's own price, while *cross-elasticity of demand* considers the price of an alternative good. If we denote price, income, and cross-elasticity of demand respectively as η_P, η_I, and η_{PA}, then we can define these with reference to equation (5.2) as follows:

$$\eta_P = (\partial Q_D / \partial P) / (P/Q) \tag{5.3}$$

$$\eta_I = (\partial Q_D / \partial I) / (I/Q) \tag{5.4}$$

$$\eta_{PA} = (\partial Q_D / \partial P_A) / (P_A/Q) \tag{5.5}$$

Business managers of firms that produce goods are specifically interested in the relationship between price elasticity of demand and total revenue. This is illustrated in figure 5.4, where we observe that total revenue is a maximum when the elasticity of demand equals one. So if the observed elasticity of demand is less than one, then managers will be inclined to raise the price of the good and they will be inclined to do the reverse should the observed elasticity be larger than one. The challenge for us is to relate this principle to environmental issues.

In this context, it would be useful to develop some elasticity concepts based on the environmental attributes of a good. This is because, as we have seen, the elasticity concept provides useful information for a business manager for pricing and revenue considerations. So if a manager is contemplating offering a new product on the market and its main feature is going to be its *environment friendliness*, then a concept such as the *elasticity of environment friendliness* (EEF) could become pertinent. That is, the higher the value of EEF, the higher the potential for the manager to capture a share of the market.

An example, though a hypothetical one, may help explain the relevance of the EEF. In almost all households (at least in the developed world) items of crockery are ceramic goods. The manufacture of these goods involves depletive extraction processes (for example sand-mining) and high energy-consuming production processes (for example glass-blowing). We should of course note that the manager's proposed crockery goods are those used daily and not those that possess an intrinsically high value owing to tradition, such as items that are stored away in household cabinets for special occasions such as the Royal Copenhagen, the Rosenthal, or the Royal Doulton. Suppose that the manager's substitute goods are cups, saucers, and plates made from the shell of the humble coconut. The manager

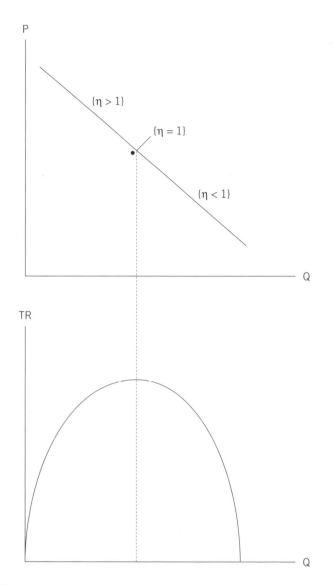

Figure 5.4   The relationship between price elasticity of demand and revenue

thinks that a production process can be improvised so that many of the attributes of the new good do not differ from those of the ceramic good; for example appearance, and the taste of the food or beverage when consumed using the new good. The main distinguishing environment-friendly attribute of the new good, however, is that it originates from a sustainable source. The business manager could of course be guided by other considerations. Most coconut plantations are found in developing countries where labour is plentiful and cheap and an investment in plant and structures may not be as high as in developed countries. But the manager also needs to have some knowledge of whether he or she would be able to capture a share of the ceramic goods market. When environment-friendliness is the significant attribute of the new good, the EEF could prove to be useful.

In the first instance, it is possible to gauge the size of the EEF by recourse to the cross-elasticity of demand. But here, the more influential parameter would be the price of the new good and not the environment-friendly attributes of that good. Suppose that we are able to construct an index of environment friendliness (IEF) for both the goods through lifecycle analysis, which explores the life history of a good from its cradle to its ultimate demise. For example this can involve aggregating the sum of toxic emissions in the lifecycle of the goods and using the inverse of this aggregate as an index. So the EEF can be defined as the percentage change in quantity demanded in response to a one per cent change in IEF. That is:

$$\eta_{EEF} = \{\partial Q_D / \partial(IEF)\} / \{(IEF)/Q\} \tag{5.6}$$

As we shall observe later, much work has been done on the formulation of environmental indexes—especially pollution indexes—and in some cases methods have been proposed to translate these indices into abatement costs.

Concluding remarks

In this chapter, we have attempted to show how some revisions to the standard theory of consumer demand could assist decision-makers in formulating strategies that are considerate of the natural environment. The first was the deviation between WTP and WTA. If this deviation is marked in the context of an environment-friendly good that is proposed by a business manager, then there is room for the manager to perform further market surveys to study the potential for introducing the good. The manager could also ascertain the potential for his or her good through an understanding of the EEF. The higher the EEF, the better the market prospect, and the manager could then examine the cross-price elasticity. Once the decision to produce the environment-friendly good has been made, the manager can promote demand for the good through advertisements and various marketing strategies that are aimed at improving the environmental awareness of the consumers.

Box 5.2 Elasticity of environment friendliness

Monsanto, the major multinational firm that specialises in the sale of agricultural commodities, has developed a new EF product—the New Leaf Potato. Of course there is nothing new about the humble potato that millions of people across the world chew on for their daily dose of carbohydrate. However, the New Leaf Potato is a bio-engineered product. This new species of potato can defend itself against species of beetles that prey on the leaf of the potato plant. This new product will remove the need for aerial spraying with chemicals to control the potato beetle. This in turn will prevent the accumulation of chemical residues in soils and in waterways.

Let us now consider a hypothetical example. Suppose that farmer Keller from Germany decides to buy the bio-engineered seed potatoes from Monsanto—Germans are heavy potato eaters. But will his consumers be impressed by the fact that there will be less chemical spraying on his property? Given the level of environmental awareness among German consumers, the answer is probably yes. Mr Keller's promotional campaign states that his potatoes will not only help the environment but will also reduce the national import bill. The control chemical is currently imported into Germany.

At the moment Mr Keller sells his produce directly to major German retailers such as Aldi and Karsdat. Having reviewed Mr Keller's spiel about the enhanced environmental attribute of his good, all retailers have decided to increase their purchases from Mr Keller. This means that Mr Keller will enjoy a 10 per cent increase in sales. Suppose that we nominate the cost savings in chemical spraying to be a measure of IEF. If the need for chemical spraying is completely eliminated, the change in IEF is 100 per cent. In this hypothetical example, the EEF is equal to 0.1. This information would prompt Mr Keller to review his pricing strategy.

REVIEW QUESTIONS

1 Critically evaluate the theoretical premise presented in this chapter to explain the distinction between WTA and WTP.
2 Do you agree with the statement that firms can succeed in promoting goods with environment-friendly attributes if the promotion includes a price reduction?
3 Provide a few examples where the estimation of the elasticity of environment friendliness (EEF) would prove useful. How could this EEF be quantified?

Online Discussion

Elasticity of Environmental Friendliness –
is this a pertinent concept?

6

Production, Costs, Supply, and the Environment

In the previous chapter, we illustrated how a decision-maker's perception of consumer demand would change when environmental considerations are internalised into the pertinent frameworks that explain consumer demand. This altered perception of demand alone is not sufficient. In this chapter we illustrate how the decision-maker would have to re-examine his or her own production and supply capability in the context of environmental considerations.

The literature on production and supply in microeconomics proceeds as follows. It commences with the description of a two-factor (labour and capital) production function that displays diminishing marginal returns. The production function is then used in the demonstration of the concept of isoquants and the cost-minimising combination of factor inputs. The production function also serves as the basis for the demonstration of specific relationships between short-run average costs and marginal costs with the latter being the supply curve in the short run. Finally, the production function and the isoquants also provide the basis for explaining the role of technology and economies of scale. The environmental economists' contention is that environmental quality is also a factor of production. We refer to this as natural capital. If a decision-maker were to recognise this additional factor of production, then a sequence of implications follows. First the nature and shape of the production function itself is altered. This leads to the isoquants being altered, which in turn has implications for the selection of factor inputs. Finally, the cost curves also could take on a different shape and this has implications for a producer's supply response and views on the role of technology and scale effects. We will consider each of these in the sections that follow.

At this point we need to make one important qualification. In the discussion that follows we assume that a producer has full control of the natural capital that he or she is using in the production process. In reality this is not always true, because the quantity and quality of natural capital that a producer uses can be affected by the decisions of other players in society. This is because in most instances natural capital does not have a well-defined set of property rights. Nevertheless, pursuing the discussion in the way we have presented in this chapter—that is, assuming that the producer has full control of natural capital—paves the way for us to consider the context where this control is absent. However, we will postpone a discussion of this context until chapters 8 and 9, where we will consider the role of governments and institutions.

The production function

The standard production function that is introduced in most texts is:

$$Y = f(KM, L) \tag{6.1}$$

where KM represents manufactured capital and L stands for labour.

In figure 6.1, we illustrate this function with reference to one factor while the other is held constant. Following the law of diminishing marginal returns, it is customary to distinguish between three zones of production. Increases in factor

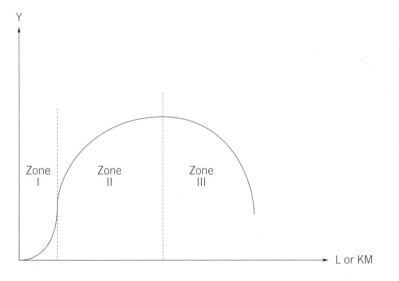

Figure 6.1 The standard production function and zones of production

quantities are rewarded by:

- increases in output at an increasing rate in the first zone
- increases in output at a diminishing rate in the second zone
- decreases in output in the third zone.

The second zone is often described as the rational zone, in that it is in this zone that the producer will choose his or her output decision.

Suppose now that we introduce natural capital also as a factor of production. We denote this as KN, and rewrite equation (6.1) as:

$$Y = g(KM, L, KN) \tag{6.2}$$

We can envisage KN as an explicit factor in a variety of business contexts. For example on the factory floor of a production line, where we usually think of KM and L as the only factors, we can also recognise indoor air quality as an important factor that represents KN. Should indoor air quality deteriorate, then it will affect the performance of L and may also affect the functioning of KM. In the case of a farm-firm, KN is represented by soil quality as well as the quality of the air and water that service farming. For a wine maker, KN is the water that is drawn from a nearby river system. We could of course think of numerous other examples in this vein. However, the production function for $\{Y = g_1(KN)$ where KM and L are held constant$\}$ can take on a distinct shape, as we illustrate in figure 6.2. This function displays the following features:

1   First the function has a fixed domain. It is for this reason that the curve describing the changes in output has a bold circle at the end of it. The fixed domain exists because there is a fixed upper limit for KN as dictated by the laws of nature. For example in the case of air quality, the law of nature dictates that we cannot have more than 20 per cent oxygen in the air. Likewise, for soil quality in a specific region, there is a maximum limit for attributes such as percentage of organic matter and porosity. Water quality is also limited by the maximum level for the percentage of dissolved oxygen.

2   Beyond a certain threshold level of KN, denoted by KN_T in figure 6.2, increases in KN have no effect on the size of Y. We can explain this in the reverse order as well. That is, if we let KN deteriorate from its maximum upper limit, KN_U, output remains unaffected until KN reduces to KN_T. When KN reduces to below KN_T, then output begins to fall. This can be explained by the fact that KN is endowed with the characteristic of assimilative ability. So the range between KN_T and KN_U can be described as the region of assimilative ability. At the same time, we can also observe another threshold value for KN, namely KN_0. As the size of KN reduces to KN_0, output approaches zero. For convenience, we will refer to KN_T as the upper threshold, and KN_0 as the lower threshold.

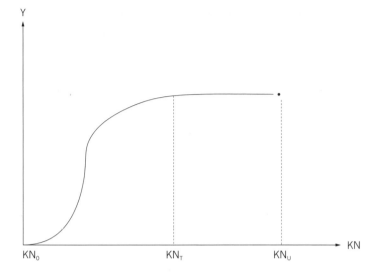

Figure 6.2 The production function relating to KN

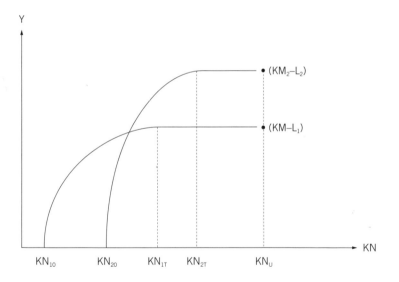

Figure 6.3 Family of production functions and increasing fragility of KN

3   In figure 6.3, we show a family of curves for $\{Y = g_1(KN)\}$, where KM and L are held constant at different levels. We try to show that as the level of KM and L gets higher, the size of Y becomes bigger, but the range of assimilative ability becomes smaller. That is, at higher intensities of KM–L, we expect KN to

become increasingly fragile. This is consistent with the second law of thermo-dynamics we observed in chapter 2, and we can easily relate this feature of increasing fragility of KN to a variety of business contexts, regardless of whether it is a farm-firm or a factory floor with a production line. Further, the gradient of the function is shown to be steeper at higher levels of KM–L than at lower levels. This implies that when KN deteriorates to a level below its threshold level, then the fall in output is much faster at higher KM–L levels than at lower levels. Also note that the upper and the lower threshold values for KN move progressively to the right as we move to higher levels of KM–L.

If the business manager begins to recognise this type of production function (as shown in figures 6.2 and 6.3), then we would expect to observe changes in his or her business strategies. Increasing the intensities of KM–L to achieve higher levels of output may not always prove to be prudent, because with the increasing fragility of KN, the manager can easily lose his or her production capability. These pressures become more explicit when we consider substitution behaviour within the framework of isoquants.

Isoquants, substitution, and input mixes

An isoquant is in fact a contour taken out of a multi-dimensional production surface. For purposes of illustrative convenience, we shall confine ourselves to a

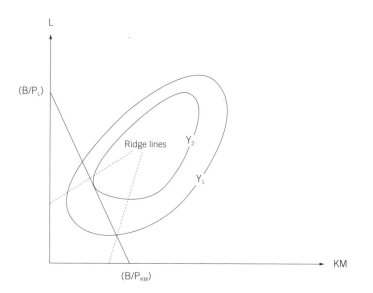

Figure 6.4   Isoquants in standard theory

two-dimensional surface. In figure 6.4, we present the isoquants that correspond to equation (6.1). Each isoquant describes a locus of points along which output remains constant for various mixes of the inputs. The region of an isoquant that is convex to the origin corresponds to zones I and II in figure 6.1. In this region there is positive substitutability between inputs. The region of the isoquant that is concave to the origin corresponds to zone III in figure 6.1. In this region, we need to increase both L and K together to maintain a certain level of output. The ridge lines are usually constructed to separate the zones in terms of substitutability.

As one would expect, a firm would confine its decision-making to the region of the isoquant that is convex to the origin. A manager who is restricted to a fixed budget of \$B, and facing input prices P_{KM} and P_L, would optimise the selection of inputs at the point where the budget line—defined by $([P_{KM} * KM) + (P_L * L) = B$—is tangent to the highest attainable isoquant. This is illustrated in figure 6.4. A familiar result in microeconomics is that, at the point of tangency, the marginal rate of substitution that is the ratio of marginal products equals the ratio of input prices. That is:

$$MRS_{L,KM} \quad = \quad (MP_L/MP_{KM}) \quad = \quad (P_L/P_{KM}) \tag{6.3}$$

As the firm's budget improves, the manager can move to a higher isoquant and thereby trace an expansion path.

We need to now consider how the above explanation will change when the business manager begins to recognise KN as also a factor of production. Because we want keep our illustration to a two-dimensional level, we shall regard manufactured capital and labour as a composite factor input (KM–L). So equation (6.2) will be modified to read as $\{Y = g_1[(KM–L), KN]\}$. The isoquants that describe this production function are shown in figure 6.5. Note that there is a correspondence between these isoquants and the production function shown in figure 6.3.

To explain these isoquants, let us go back to the example of the factory floor with a production line where KN is represented by indoor air quality. Consider the isoquant labelled I_1I_1. Substitutability between KN and (KM–L) begins when the size of KN falls below its threshold level KN_{1T}. That is, there is no substitutability between factors in the region of assimilative ability that is defined by $\{KN_{1T} < KN < KN_U\}$. Substitutability between KN and (KM–L) also ceases at some extremely low level of KN, namely KN_{10}, where output approaches zero. At this level (KN_{10}) the isoquant straightens up, implying that at this point we would need infinite quantities of (KM–L) to replace KN. Further, as we move to a higher isoquant, say I_2I_2, we note that the region of substitutability becomes smaller and threshold quantities KN_{2T} and KN_{20} begin to appear sooner.

One of the questions that emerges is—where do we place the ridge lines? One ridge line can be clearly constructed, and this will be the one that connects the lower threshold levels of KN at which substitutability fails: that is $(KN_{10}, KN_{20},$

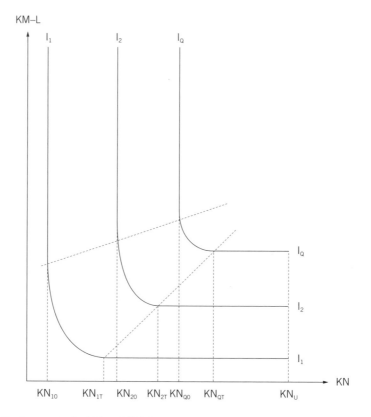

Figure 6.5 Isoquants for KN and KM–L

..., KN_{Q0}). If we place the other ridge line to connect the upper threshold quantities (KN_{1T}, KN_{2T}, ..., KN_{QT}), we imply that it is rational for the producer to let KN degrade and input choices be confined to the region of substitution only. Alternatively, it is possible to describe the next ridge line as a vertical line representing the fixed domain for KN. This would ensure that conserving KN, or rather exploiting the assimilative ability of KN, is also an input option for the producer.

However, we are unable to resolve one major issue of strategy that gets resolved with other factors of production. This is the determination of the input mix, as illustrated in figure 6.4. This difficulty is of course primarily due to the absence of a price for KN. But it is certainly clear that a producer should not choose a production strategy that traces a path of the lower threshold values of KN (KN_{10}, KN_{20}, ..., KN_{Q0}). This is because the producer runs the risk of losing his or her production capability. Alternatively, the producer may nominate the upper threshold levels (KN_{1T}, KN_{2T}, ..., KN_{QT}) as the basis for an expansion path, but needs to be cautious of the fact that the gap between the upper and lower thresholds becomes progressively narrower as the intensity of KM–L gets higher.

Being on the upper threshold, however, is not too far from selecting a strategy for KN that places it within the region of assimilative ability. Further, if the price of KN is deemed to be zero, the budget line is always horizontal, and the input strategy for KN would always be within the region of assimilative ability. One major implication of the framework we have considered here concerns the relationship between governments and firms. If governments wished to force producers to operate within the region of assimilative ability and higher stability (that is, lower levels of intensity for KM–L), then they would have to place taxes on KM–L so that the budget line shifted to a stable level of output. The limit on substitutability when KN has been rendered fragile at high intensities of KM–L has implications for the theory of costs, and we will consider this next.

Analysis of costs

As indicated, the theory of costs has its foundations laid on the underlying production function. The total cost (TC) function (which comprises fixed costs and variable costs) is usually described as one that increases:

- initially at an increasing rate
- then at a slower rate
- then at an increasing rate.

In decision-making, apart from TCs, managers are also interested the shape and properties of average costs (AC) and marginal costs (MC). While TC is the sum of all relevant costs during a production period, AC is the cost per unit of output produced: that is, (TC/Q) where Q is the quantity of output. MC is the TC of producing one extra unit of the good and is usually denoted by (dTC/dQ).

For decision-making, managers will usually compare the TC with total revenue (TR) and select a production quantity that maximises the departure between TC and TR functions. The relationship between AC and MC also plays an important part in decision-making. As Q increases, both AC and MC initially decrease and then increase beyond a certain value of Q, which represents the shift from the first to the second zone of the production function. However, there is one important property of cost functions that should be noted. That is, *when costs are falling, AC is greater than MC, and when costs are increasing MC is greater than AC*. This is illustrated in figure 6.6 (p. 68). It would be irrational for a decision-maker to produce in the region where AC is greater than MC, because price is equated to MC, and hence in this region TR falls short of TC. Therefore, decision-makers will choose their production strategies in the ascending region of the

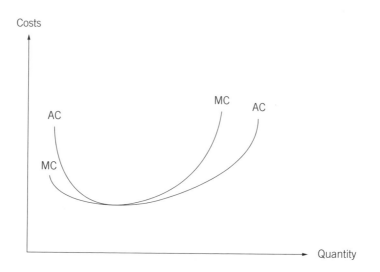

Figure 6.6   Average and marginal cost in standard theory

cost functions. It ~~is for this reason that the~~ *supply curve in microeconomics is often described as the ascending segment of the MC curve.*

Consider now the implications of the conceptualisation we have introduced above with KN as an additional factor of production. If a producer were to increase output by increasing the quantities of L and KM, he or she would be

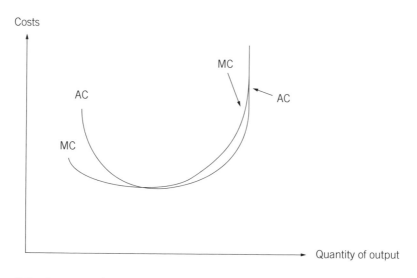

Figure 6.7   Average and marginal cost that tends to infinity

shifting upwards from one production schedule to another, as in figure 6.3, where by the fragility of KN progressively increases. At some level of KM–L, output capability completely disappears due to the complete break down of KN. Hence we can describe the cost curves very much the same way as it is done in standard texts, but with one exception. This is that the producer's costs will tend towards infinity, at the point where KN breaks down completely. Hence the supply curve of a business will be upward sloping as expected in theory, but, turning vertical at some limiting point. In other words, recognising KN as a factor of production makes the supply curve of most goods look like that of a non-renewable resource in natural resource economics.

Implications

In this chapter we have considered three concepts that economists employ to illustrate how producers can formulate production strategies, namely the production function, isoquants, and cost functions. This formulation is of course subject to the provision that other pertinent management variables such as market potential and competitive advantage have been considered. The production strategy will basically involve the quantity of production, the allocation of resources, the choice of an expansion path, and response to prices and relevant market signals. In the absence of environmental considerations, we can summarise the role of the three concepts as follows:

1 The production function assists with the selection of input quantities and identifies the range within which output decisions can be made.

2 The isoquants provide the same information as in (1), but at the same time enable the producer to trace out an expansion path when budgetary conditions improve. This can be of assistance with medium- to long-term planning.

3 The cost curves assist with information on how the producer could respond to prices. Note that the upward sloping segment of the marginal cost curve is the supply curve. So, in theory, if prices rise the producer can incur higher marginal costs and raise production. In practice, however, producers cannot vary production quantities to match every price variation. Yet the cost curves at least do provide information on the range over which supply responses can be varied. For example if prices are expected to fall below a certain level, say where AC exceeds MC, then the producer has to seriously consider plans to exit from the industry.

When environmental capital is introduced as an additional input, we find that there are some important changes. First, it is not so straightforward to pursue an expansion path when the budgetary conditions improve. This is because higher levels of the non-environmental factor intensities are associated with increased fragility of the environmental factors. This also implies that the benefits from the economies of scale that are observed with higher production quantities may be overstated.

The main implication of the conceptualisation that we have presented here is that an additional limit to the production capability and the response capability is induced by KN. Note that the producer's endowment with respect to KM and L also represent a limit on the production and response capabilities, at least in the short run. Therefore it is pertinent for the manager to determine which limit is enforced first; that is, whether the limit enforced by KN emerges first or the limit enforced by KM–L emerges first. Should the limit enforced by KM–L emerge first, then there is scope for expansion up to the limit imposed by KN.

It is customary to explain the role of technology in production through either upward shifts in the production function or fixed levels of factor utilisation. This manifests in the proximal bunching of the isoquants (compared to isoquants being spaced apart). When KN becomes a limiting factor, it is possible for the producer to explore whether it is worthwhile to focus on developing a technology or production method that would alleviate the limitations imposed by KN. This can involve searching for a technology that extends the region of assimilative capacity on the production function, or increases the range of substitutability on the isoquant. The appropriateness of seeking such a technological innovation would of course be determined by a wider range of factors such as price expectations, competitiveness in the market and demand expectations. Should such variables signal that the innovation is appropriate, then KN becomes an instrument that contributes to the manager's competitive advantage.

But how real is KN on the production line? As indicated above, we can observe direct relevance in agriculture and forestry. The quality of farm products derived from quality soils is bound to be superior to that of those drawn from contaminated soils. Performance of workers on worksites that have higher aesthetic qualities is also bound to be higher than it is on sites that display poor aesthetic qualities. It is possible to argue that the output per worker in a steel plant in Australia, where all emissions are strictly filtered, is higher than in a similar plant in Korea, where the emissions are not controlled, despite lower wages in Korea. Thus, in spite of higher costs, it is possible for Australian steel to remain competitive.

However, in many instances, the producer does not have complete control over the deterioration of KN. For example the cyanide spill in Hungary by Esmeralda, the Union Carbide accident in Bhopal, and several others illustrate this fact. It is in this context that we need to consider the role of the government and regulatory bodies, and this is taken up in chapter 9.

The conceptual tools presented thus far enable us now to deal with the valuation of environmental goods and services as well as with the formulation of policy frameworks. These are considered in turn in the chapters that follow.

Box 6.1 The Murray River—a case of fragile KN

Perhaps the best example to illustrate the fragility of KN in Australia is the Murray River (the fourth-longest river in the world). For a moment, suppose that there is just one isolated and very small farm property in the entire area that spans the Murray River Basin. This is indeed a purely hypothetical scenario. But in this abstract scenario, the intensity of KM–L would be so low that the Murray River would not be threatened with the problem of degradation and fragility. However, in reality, as reported in the Australian Broadcasting Corporation's *Landline* program (24 September 2000), the highly fragile state of the Murray River has been due to the existence of:

- many farms and associated activities such as land clearing
- several secondary industries such as canneries, wineries and fruit juice manufacturers.

This level of activity in the Murray Basin points to a very high level of intensity for KM–L. Because the Murray River is an explicit input in the production function of the various firms on the Basin, either in terms of providing water or acting as a receptacle for run-off and discharges, efforts have been taken to restore and stabilise the river. One such effort was pioneered around 1986 by Mr John Moore, the director of one of Australia's largest fruit juice manufacturing companies, in the town of Loxton in South Australia. Mr Moore's initiative was to plant some 250,000 eucalypt trees on a land area of 100 hectares in the river basin, and then pump the wastes from the fruit juice factory together with the effluent from the town into this land area. These wastes, which would otherwise have ended up in the Murray, became useful nutrients for the eucalypts. In turn, the eucalypts were capable of reducing salinity in the river system. The project was of course initially successful because, as stated in the Landline program: 'Eucalypts like river red gum and flooded gum are proved water users; they act like a set of lungs pushing water out of the ground and into the atmosphere at twice the rate of passive evaporation.' But unfortunately, around 1996 Mr Moore's company went into receivership and the effluent from Loxton was not sufficient to maintain the trees. The eucalypts in the land area that has become known as 'Moore's Woodlot' have begun to die.

Source: http://www.abc.net.au/landline/stories/s184313.htm

REVIEW QUESTIONS

1 Consider the following sectors of the economy: tourism, manufacturing, agriculture, fisheries, and mining. Illustrate how firms in each of these sectors can conceptualise the existence of a production function: $Y = g(KM, L, KN)$.
2 In the examples chosen for question 1, illustrate the substitutability between KM–L and KN.
3 In the same examples chosen above, describe how firms could utilise KN more efficiently.

7

Valuation of Environmental Goods and Services[4]

As indicated in chapters 3 and 4, the valuation of environmental goods and services is central to the correction of market failure. Due to improved environmental awareness, many recent commercial as well as public decisions seem to hinge on values ascribed to environmental outcomes. Thus the valuation of environmental goods and services has become increasingly important for policy-makers. Although the applications of methods for environmental valuation date back as far as the 1930s, concerted efforts to use these values in policy formulation is a recent development. The object of this chapter is to present a concise review of the various valuation methods that are pertinent to environmental goods and services. As indicated, the conceptual tools considered in the previous chapters would now become handy for a better appreciation of some of these methods. For convenience, we use the term 'environmental quality' (EQ) to denote the various goods and services that are recognised by decision-makers in a wide range of contexts.

The basis for valuation

The literature on environmental economics considers three types of values, namely *existence value, option value,* and *use value.* The first of these, existence value, is the value that is attributed by individuals to the mere existence of an EQ and has no bearing whatsoever on their use of this item. Option value refers specifically to the preservation of environmental options that have no current value but potential future value. Hence it captures the right that individuals

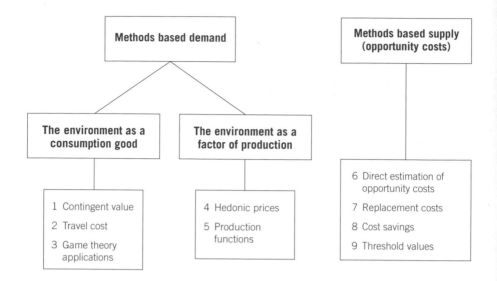

Figure 7.1   An overview of the valuation methods for environmental outcomes

would like to exercise in the future in terms of enjoying the services of EQ. Use value deals with various forms of EQ that are in current use. As illustrated, most methodological developments are centred on use value.

Following standard microeconomic theory, the value of a good or service is defined by the area below its demand curve, and/or the area below the supply (marginal cost) curve. The former measures value in terms of either willingness to pay (WTP) or willingness to accept (WTA), while the latter equates value to the opportunity costs of providing the good.

Following this conceptual premise, the methods of environmental valuation can be classified into two broad categories, as shown in figure 7.1. As indicated below, these methods attempt to elicit the value of EQ by either improvising markets or making use of actual transactions that occur in related markets. Each method of valuation is considered in turn under the relevant heading, with reference to specific examples.

Methods based on demand or WTP—EQ as a consumption good

Most of the applications that have been reported in this category attempt to value EQ as a good that is consumed. That is, EQ enters directly into a consumption utility function. Hence, either direct or indirect estimation of consumption utilities is an essential feature of these methods.

Contingent valuation method (CVM)

The contingent valuation method has been used mostly to estimate the value of wilderness areas in terms of either preservation or the losses inflicted by damaging economic activities. It is claimed to be relevant when the wilderness area in question has several important environmental features. This does not imply that this method can readily capture the value of all features, but rather the implication is that the other methods of valuation that are described below may not be applicable. The method has been applied widely, but not without controversy; for example the valuation of the effects of the Exxon-Valdez oil spill in Alaska (Carson et al. 1992); the value of preserving the Kakadu National Park in Australia (Wallace, 1992); rural water supply decisions in Nigeria (Whittington, Okorafor, & Okore 1990); and preserving the Kariba lake shore in Zambia (Thampapillai, Malcka, & Milimo 1992).

In its simplest application, to estimate the value of environmental preservation, the method involves the direct questioning of a sample of individuals on their WTP for preservation. As a first step, the person being questioned is made aware of the characteristics of EQ and is asked whether he or she would be willing to pay a certain amount, say $500 per year, for EQ. If the response to this initial question is 'yes', then no further questions need be asked, and the person's WTP is recorded as $500. However, if the response is negative, then the person is asked whether he or she would be willing to pay either more or less than $500. If the answer is 'more than $500', then the person is bidded up until he or she says 'no more'. Alternatively, if the answer is 'less than $500', then the person is bidded down until he or she says 'no less'. By repeating this method in this manner with various individuals, an average value of WTP can be derived for the sample that is questioned. The method derives its name from the fact that the value of WTP that is derived for each person is contingent upon the initial information that has been provided. The value of EQ can be determined by multiplying the average value of WTP by an appropriate population factor.

However, several issues emerge in the context of applying the CV method. These pertain to the types of biases that can prevail among those carrying out the survey, as well as among those responding to it (Pearce & Turner 1990), and the lack of consistency between hypothetical transactions and actual economic commitments (Neill et al. 1993). Response biases were evident in the survey of individuals in Lusaka for the estimation of WTP for the preservation of the Kariba lake shore (Thampapillai, Maleka, & Milimo 1992). While at one extreme some people were willing to contribute in excess of fifty per cent of their income for preservation, at the other extreme there were people who were willing to contribute none. Further, some argue that the public good characteristics of environmental goods and services could prompt respondents to overstate their WTP, while others argue the reverse by recourse to experimental economics (for example see Cummings & Harrison 1992).

Knetsch (1994) provides a wide range of examples to clearly illustrate two significant difficulties with the CV method. These are *anchoring* and *embedding*. Anchoring is evidenced from the observation that most responses are centred on the initial bid value that is proposed. In some instances, when the individuals were given unrelated information containing irrelevant numerical values, a spurious correlation between these irrelevant numbers and the average bid value was observed. Embedding is illustrated by the fact that most individuals set aside a certain amount of their income for altruistic purposes. The value assigned for EQ is embedded in the WTP or WTA for altruism as such. Hence, when questions are not properly formulated, it is the value for altruism rather than an individual component of EQ that is ascertained.

Despite the above difficulties, CVM continues to be a method that is popular because of its ease of application and its applicability to a diverse set of contexts. Some specific advances have been made by Frykblom (1997) in terms of calibrating the responses in CVM surveys for consistency. Further, some of the inconsistencies and disparities in CVM responses have been minimised by recourse to a dichotomous (binomial) response method. Here the larger sample is divided into smaller subsamples, each of equal size. The response sought is that of a 'yes' or 'no'. The first subsample is asked of its WTP or WTA in excess of a trivially small amount, such as $1. This amount is gradually increased across the subsamples and the final sample is exposed to a very large amount, say $5000. It is now possible to search for the relationship between the proportion of persons saying 'yes' in each subsample and the WTP or WTA amount that was offered to each subsample. As one would expect, this would be a downward sloping curve and would be a proxy for the demand curve.

Travel-cost method (TCM)

The travel-cost method is applicable to the valuation of sites that have recreational potential. The underlying assumption is that the value of EQ is equal to the value of recreational benefits that are provided by the site. The method has been applied and illustrated widely; for example Knetsch (1963, 1964), Clawson and Knetsch (1966), Sinden (1973), Sinden and Worrell (1979), and Pearce and Turner (1990). The method in its simplest form involves asking a sample of visitors (recreationists) at a given site two questions. These are: 'How far do you have to travel to visit this site?' and 'How often do you visit this site?'

By grouping the respondents into areas that are equidistant from the site, it is possible to find for each group an estimate of the average number of visits per year. The 'distance travelled' is then translated into a cost that includes travel and other related recreational expenditures. This cost, which is referred to as the 'travel-cost', is taken as a proxy for the price of recreation, while the average number of visits per year is taken as a proxy for the quantity of recreation.

Because those who live closer to the park visit the site more frequently than those who reside further away, the relationship between the travel-cost and the number of visits is usually inverse, as shown in figure 7.2. This relationship is in fact the demand for recreation, and by virtue of the assumptions that have been made it is also the demand for the environment. The value of EQ at the site can now be found by measuring the area below the demand curve.

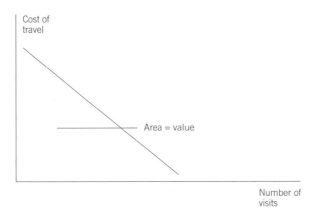

Figure 7.2   The demand for the EQ by the TCM

In some instances, the travel-cost method can be modified to include some features of the CVM method. For example it is possible to estimate the value of wilderness areas that offer facilities for tourism by bidding respondents with tour package prices, in terms of the number of visits they are likely to make per year. That is, the main question asked of respondents is: 'How many times per year will you visit the wilderness area for X days, if you have to pay $Y?'

Tisdell (1991) provides a comprehensive list of deficiencies that are associated with the method. These include the non-homogeneity of population attributes, multiple purposes that can be associated with recreational visits, and the fact that the economic value of a natural area cannot be determined by the number of visits alone. Further, given that this method is applicable only to those sites that have recreational facilities, it is not readily relevant for the valuation of remote wilderness areas such as those contested for mining leases.

Game theory method

The salient features of the game theory method are as follows. Each individual is subjected to a game theoretic model, from which the person's indifference map is derived. This map is then used to derive the person's demand curve for preserving an environment. By deriving a sample of individual demand curves and then adding them, an aggregate demand curve can be derived. The value of the environmental good or service is then measured by the area below the aggregate demand curve. Although a variety of game theoretic models can be used, the Ramsey model has been used by Sinden (1974), Sinden and Wyckoff (1976), and Thampapillai (1985).

For example Thampapillai's (1985) application of the Ramsey model involved the elicitation of a surrogate decision-maker's indifference map for two policy goals, namely income maximisation (IM) and environmental quality (EQ). The application of the Ramsey model to the individual resulted in the display of a family of utility curves for his utility towards EQ, as illustrated in figure 7.3A. Each utility curve in this family of curves explains the variations in the individual's utility for EQ at a fixed level of IM. For example in the lower-most curve of figure 7.3A IM is fixed at IM_1. Consider now the cross-section across the family of utility curves at utility level U_1. The points on this cross-section represent various combinations of IM and EQ that give the same utility of U_1 utiles. These points then form the basis

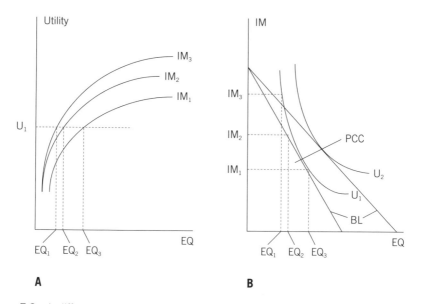

Figure 7.3 Indifference map from a game theory model. A—indifference map of an individual; B—the resulting utility curves

for an indifference curve as shown in figure 7.3B. Several cross-sections such as AB enable the demonstration of the individual's indifference map.

Demand for EQ can now be derived by nominating a budget and an appropriate price for IM, say the return per dollar investment. By arbitrarily setting a range of prices for EQ, a series of budget lines (BL), as illustrated in figure 7.3B, can be derived. Following standard microeconomic theory, the various points of tangency of the budget line with the indifference curves result in a price consumption curve (PCC), and from this curve the demand curve for EQ can be derived. The main difficulty with this method is that it deals with hypothetical transactions, and is hence inevitably inadequate in portraying true economic commitment. Besides, the method is also time-consuming.

Methods based on demand—EQ as an input in production

In all methods reported in this category, a production function is either explicit or implicit and the demand for EQ is derived in terms of the concept of marginal value product. Hence, the value of EQ enters consumption utilities indirectly through goods and services that rely on EQ for their provision. Therefore the demand for EQ is based on the concept of the marginal value product.

Hedonic prices

This method is usually applied to estimate the loss in EQ caused by the polluting activities of firms. Such loss is often reflected in the loss in property value (PV). Hence the method approximates the value of EQ to the differentials in PV. In most applications, a relationship between PV and EQ is estimated through a regression analysis. In some instances, a step-wise regression procedure is used to narrow the relationship between PV and EQ. Should the relationship between PV and EQ be significant, then the demand for EQ is defined in terms of the rate of change in PV with respect to EQ. For example suppose that {PV = f(EQ)} displays diminishing marginal returns. Then the derived demand for EQ will be a downward-sloping function, namely {(dPV/dEQ)= g(EQ)}; that is, the marginal value product.

While it is true that changes in EQ can prompt variations in PVs, the hedonic method cannot be applied over a wide geographic area, and has to be confined to a closed geographic area such as the vicinity of a polluting industry. Tisdell (1991) argues that differentials in PVs can at times be poor indicators of the value of preserved natural areas. This is due to PVs closer to the natural areas being sometimes lowered due to restrictions imposed on the property owners to protect the attributes of the natural area; for example limitations on the methods of waste disposal and the holding of domestic pets.

Isoquants involving EQ

With isoquants involving EQ, the principles are similar to those outlined above. However, as explained in chapter 6, when a production function with EQ as an input among others—say labour (L) and capital (K)—is considered, the derivation of the demand for EQ is not as straightforward as when EQ is the only input. For example in several recent studies, agricultural output (Y) has been defined as a function of EQ (measured in terms of soil conservation) (Walker 1982; Walker & Young 1986; van Vuuren 1986; Sinden & Yapp 1987; and Sinden & King 1988). So given a production function, as suggested in chapter 6, involving L and K along with EQ, that is:

$$Y = m(L-K, EQ)$$ (7.1)

the demand for EQ will not be the partial derivative $(\partial Y/\partial EQ)$. Instead, it will be a function based on this partial derivative where shifts in the marginal value

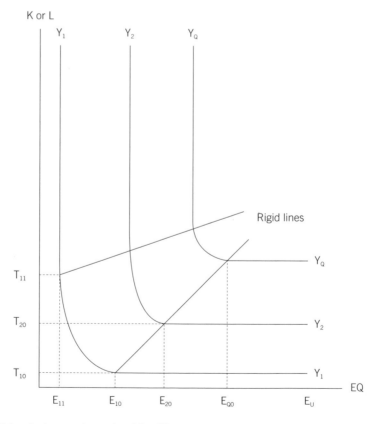

Figure 7.4   An isoquant map involving EQ

product are identified. However, given that it is difficult to identify such shifts, an alternative procedure involving the isoquants derived from the production function may be appropriate. The nature of the isoquants that are likely to prevail between EQ and a traditional input such as L or K was illustrated in chapter 6 and is reproduced here in figure 7.4. E_U is the upper limit for the value of EQ as dictated by the laws of nature; for example perfect air quality cannot have more than twenty per cent oxygen.

To recapitulate the explanation given in chapter 6, consider the isoquant associated with a constant output of Y_1. As long as the quantity of the EQ input is in excess of E_{10}, Y_1 units of output can be produced with a fixed quantity of L–K, namely T_{10}. When EQ degrades to a level below E_{11} (for example due to soil erosion in agriculture), more L–K has to be substituted to maintain output at Y_1. When EQ is irreversibly degraded, say at level E_{11}, then the amount of L–K that has to be substituted for EQ tends to infinity. Hence, E_{10} and E_{11} can be defined as threshold levels of the EQ input. E_{11} represents a threshold that signals a complete breakdown of EQ, while E_{10} represents the lower limit of EQ that retains the assimilative capacity in full. If we concede that EQ has value for levels EQ $<$ E_{10}, then it is possible to estimate this value in terms of substitutability between L–K and EQ. The value in substitution approaches infinity as the amount of EQ available falls to the lower threshold.

As indicated in chapter 6, as we move to higher isoquants, that is $Y_2, ..., Y_Q$, the ridge lines trace the two thresholds. If increases in Y have to be also reconciled with retaining the assimilative capacity of EQ, then it is possible to argue that any expansion path must trace upper thresholds, that is $E_{10}, E_{20}, ..., E_{Q0}$. Suppose that a resource manager's current utilisation of inputs is (E_{10}, T_{10}) yielding an output level of Y_1. If we further suppose that the manager has a fixed resource budget, say \$B, and the price of the tangible resource (L–K) is known, namely \$W, then the price of EQ in terms of the budget is given by:

$$\{B - (W^*T_{10})\} / (E_U - E_{10})\}$$ (7.2)

As another scenario for valuation, suppose that the resource manager wishes to increase output relative to the current context of inputs (E_{10}, T_{10}) yielding an output level of Y_1. Suppose that the aspired level of output is Y_2 involving an input utilisation of (E_{20}, T_{20}). Hence the increase in output, $\Delta Y = (Y_2 - Y_1)$, involves the following changes in input utilisation: $\Delta T = (T_{20} - T_{10})$ and $\Delta EQ = (E_{20} - E_{10})$. If P, W and P_{EQ} denote the prices of output, L–K and EQ respectively, then it is plausible to argue that the manager would consider the increase in output to be feasible only when:

$$P^* \Delta Y \geq (W^* \Delta T) + (P_{EQ}^* \Delta EQ)$$ (7.3)

From 7.3 it follows that for the decision to be feasible,

$$P_{EQ} \leq \{(P^* \, \Delta Y) - (W^* \, \Delta T)\}/ \, \Delta EQ \qquad (7.4)$$

In (7.4) it is possible to define P_{EQ} as an upper limit; that is, the price of EQ cannot exceed the RHS of (7.4) if output expansion is to be feasible. The ability to perform such valuations would of course depend on the ability to measure ΔEQ adequately. Given the significant advances in environmental monitoring and measurement, the applicability of (7.4) for the valuation of EQ should be feasible. For example, should the above illustration pertain to a factory floor where EQ refers to indoor air quality, then $(E_{20} - E_{10})$ is in fact the reduction in air quality that is associated with ΔY. Similar conceptual premises have been used in the valuation of EQ under the heading of dose-response relations; for example see Mendelsohn (1992).

Methods based on opportunity costs

The methods based on opportunity costs (OCs) equate the value of environmental goods and services to the value of income benefits that have to be forgone. As indicated below, while this treatment is explicit in some methods, it is implicit in others. When these methods are used for the estimation of benefits, the contention is that the estimates represent a minimum value. That is, the benefits are *at least equal to* the opportunity costs. Generally this method has been associated with valuing the preservation of natural environments that would otherwise be used for generating some form of income.

Direct estimation of OCs

The OC of preserving a natural environment is equal to the highest net income that has to be forgone from an alternative activity. Hence, to calculate the true OC of preservation, one needs to identify all feasible alternative activities and estimate the net income of each activity. In practice, such a procedure is not feasible, and the estimation of OC is invariably limited to a select few activities. Further, the use of OCs as a basis for value is always easy when a decision has been taken. For example, when the Australian government decided to preserve the Kakadu National Park and prohibit any form of mining activity in 1993, the OC value of preservation was conveniently narrowed to the value of mining income that had to be forgone.

The concept of OC has also been central to the discussion of environmental irreversibility; for example see Krutilla and Cicchetti (1972) and Fisher and Krutilla (1985). The argument is that, should an environment be unique, then it is irreplaceable, and therefore the OC of exploiting it approaches infinity. Further, when a development activity that is intended at a given natural area is subject to technological change, then the opportunity cost of preservation is most likely to diminish over time. For example, should there be a revolution in the technology of using solar energy, then the opportunity cost of preserving natural areas with energy resource deposits such as coal and uranium, instead of mining them, would become very low. Hence, following Krutilla and Cicchetti (1972), a basic definition for the OC value of preservation is:

$$(B_d - C_d) - (B_a - C_a) \tag{7.5}$$

where:

B_d and C_d represent respectively the benefits and costs of energy from mining and B_a and C_a represent the benefits and costs of energy from an alternative technology or an alternative method.

Should B_a and B_d be equal, then the OC value of preservation is given by $(C_a - C_d)$. Hence, the argument is that when several alternative methods of income generation are explored, the cost of preserving a natural area may not turn out be far too expensive.

Box 7.1 The OC of preserving old growth forests

As an example of the OC principle for environmental decision-making, consider the controversy that emerged in 1987 when Harris-Daishowa, a wood-chip exporter, applied for a renewal of its logging licence for a period of 17 years. The main issue here was the logging of old growth forests around Eden in south-eastern New South Wales. These forests supposedly harboured some unique species. Hence the decision problem was one of choosing between preserving the forest and logging it. The argument in favour of logging centred around the claim that the export of wood-chips generated an income of approximately $40 million for Australia. On the basis of this information, one would conclude that the OC of preservation was $40 million. However, Harris-Daishowa is a fully foreign-owned enterprise, and therefore the income that Australia would have had to sacrifice in the context of preservation was not export revenue, but the tax on export revenue. For convenience, if we assume a tax rate of 30 per cent, this would have amounted to $12 million; that is, the OC preservation in terms of tax revenue forgone would have been $12 million per year.

Consider now another option, namely that of getting raw material for the chip mill from a more distant location, say from some plantation forests. Due to the higher costs associated with this option, suppose that export revenue reduces to $35 million per year. The tax earnings to Australia, if this option were feasible, would be $10.5 million. In the context of this alternative option for logging, the OC of preserving the old growth forests reduces to $1.5 million per year. In present value terms, this OC is equal to $15.67 million (assuming a 17-year period and a 7 per cent discount rate).

It may appear, had these OC values been seriously considered in 1987, that the forests around Eden might have been preserved. The then Minister for the Environment, Senator Peter Cook, recommended the renewal of the logging lease on the recommendations of a committee headed by Professor Henry Nix of the Australian National University.

Replacement costs

The replacement costs method is based on the premise that the value of an environment that is committed to development should be at least equal to the value of restoring it to its original state from the developed state. The method is reasonably relevant if the site that is proposed for development is either a forest or a wetland. Then the value of the environment can be equated to either the cost of reforestation or regeneration. Similarly, in the context of an infestation by algal blooms, the value of the environment would equal the cost of removing the blooms and treating the water supplies to restore quality levels that are suitable for consumption. The method has been applied in the context of reforestation by Water Resources Engineers (1970). When replacement includes the rehabilitation of wild life, the costs tend to be prohibitively high. No doubt the cost of replacing unique environments will be infinity.

Consider an example where the value of the environment was estimated for natural areas bearing coal deposits that were to be strip mined. The value of the EQ at any given time was equated to the sum of two items, namely:
- the OC of preservation; that is, the value of the resource left without being mined
- the cost of restoring the mined-out area to its original state.

As indicated in the method, the cost of restoration becomes a determinant of the extent of irreversibility. To illustrate this procedure, suppose that H hectares of land, bearing a uniformly distributed mineral deposit, are being considered for clearing and mining. Let V_i denote the value of the environment after the ith hectare of land has been mined. Following a definition offered by the Water Resources Engineers (1970):

$$V_i = (C_i + B_i) = (C_i) + P(q_i) \qquad (7.6)$$

where

C_i is the cost of restoring i hectares back to their original state

B_i is the income forgone by preventing the remaining (H–i) hectares from being mined

P is the profit per unit (tonne) of the mineral

q_i is the quantity of mineral remaining in (H–i) hectares.

When the mineral is uniformly distributed across the H hectares of land and the mine is a price taker, the relationship between B_i and the area mined (L_a) will be linear decreasing function. The relationship between C_i and L_a will depend on the degree of irreversibility of the environment. Should the rate of increase in C_i be higher than the rate of decrease in B_i as L_a increases, then the environment is irreversible relative to the reverse situation as illustrated in figure 7.5. Such a valuation methodology could facilitate choices between preservation and development. That is, as an initial step, miners can decide to mine those sites that are relatively reversible.

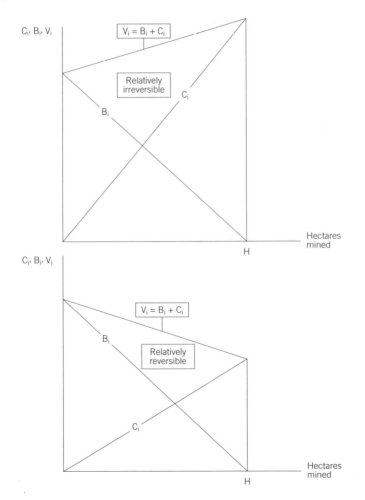

Figure 7.5 The value of the environment for varying degrees of irreversibility

Cost savings

In the cost savings method, the value of the environment is equated to the value of costs saved when the environment is left undamaged in its pristine state. An illustrative demonstration of the method is provided in Greig and Devonshire (1981), where the value of the environment is approximated to the value of tree cover. The method proceeds in two steps as follows. In the first step, a functional relationship between tree cover and salinity is derived by recording salinity levels of soil samples taken from sites with different magnitudes of tree cover. This relationship is illustrated in the right-hand section of figure 7.6. That is, the magnitude of salinity increases as the size of tree cover decreases.

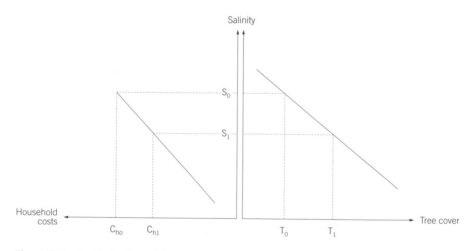

Figure 7.6   An illustration of the cost-saving method

In the second step, the study area is surveyed to find the functional relationship between household (and farm) costs and salinity. As expected, these costs increase as salinity increases. This is because, for example, increases in salinity prompt more frequent replacement of appliances and increase the expense of soil treatment on farms. The relationship between household costs and salinity is shown on the left-hand side of figure 7.6. For illustrative purposes, both relationships have been drawn as straight lines. These two relationships can now be used to estimate the value of tree cover in terms of costs that can be saved. For example consider the reduction of tree cover from T_1 to T_0 in figure 7.6. This results in an increase in salinity from S_1 to S_0, which in turn creates an increase in household expenditure from C_{h1} to C_{h0}. Hence the value of tree cover (and by assumption, the value of the environment) is approximated to the distance between C_{h1} and C_{h0}.

The concept of cost savings has been used in several contexts. For example the benefits of improving EQ by reducing pollution have been estimated in terms of savings in health and medical expenses. The benefits of improved public transport facilities can be estimated in terms of savings in road maintenance costs, as commuters may substitute public transport for private transport. Similarly, the benefits of flood mitigation can be estimated in terms of expenditure savings from the reduced number of emergency operations and charity payments.

Threshold values

Krutilla and Cicchetti (1972) introduced the concept of threshold values for the valuation of unique irreversible environments. In this context, a threshold value is defined as the minimum value of an environmental benefit (EB) stemming from the preservation of an environment in the initial year growing at a specific rate such that the present value of the EB is at least equal to the highest net present value of the alternative to preservation. That is, the threshold value (TV) is the initial year's minimum value for the EB that would render preservation just as desirable as its alternative.

If the TV can be estimated, then its validity can be subjectively appraised. The estimation of TV requires: complete data on the alternatives to preservation, and the growth rate for the environmental benefit. The latter may be determined by scientific enquiry or the use of pertinent data; for example the rate at which demand for recreation is expected to grow or the rate at which bio-diversity is expected to grow.

The subjective appraisal can be illustrated as follows. Suppose that the TV for preserving a popular beach in Sydney amounts to $500,000. One could ask the question: Is the value of nature not much more than $500,000? In this context subjective evaluation is guided by the fact that $500,000 is less than the average value of a single beachfront property in Sydney.

The definition of the TV is embedded in the simple calculus of cost–benefit analysis and can be explained as follows:

Consider two alternatives at a specific location, namely commercial development and preserving the EQ of the location. Suppose that these two alternatives are mutually exclusive in that preservation of EQ and commercial development cannot coexist.

Let the NPV from commercial development be $\$B_d$, and the value of preservation in the initial year be $\$b_{op}$. Further, suppose that the preservation benefits grow at r per cent and that the discount rate is i per cent. The NPV of preservation can be defined as:

$$= b_{op} + [\{b_{op}(1+r)\}/(1+i)] + [\{b_{op}(1+r)^2\}/(1+i)^2] + \ldots + [\{b_{op}(1+r)^T\}/(1+i)^T] \quad (7.7a)$$

$$= b_{op}[1 + \{(1+r)/(1+i)\} + \{(1+r)^2/(1+i)^2\} + \dots + \{(1+r)^T/(1+i)^T\}] \qquad (7.7b)$$

If the NPV of preservation exceeds the NPV of commercial development, then:

$$b_{op}[1 + \{(1+r)/(1+i)\} + \{(1+r)^2/(1+i)^2\} + \dots + \{(1+r)^T/(1+i)^T\}] \geq B_d \qquad (7.8a)$$

$$b_{op} \geq (B_d)/[1 + \{(1+r)/(1+i)\} + \{(1+r)^2/(1+i)^2\} + \dots + \{(1+r)^T/(1+i)^T\}] \qquad (7.8b)$$

The RHS of (7.8b) above is the TV because, if b_{op} exceeds the value estimated from the RHS, then the preservation of EQ would be the preferred option. Note that the denominator of the RHS in the last statement above is the present value of $1 growing at the same rate as the preservation benefits. Hence a formal definition of the threshold value for environmental preservation (in the context of such preservation being a mutually exclusive alternative to a well-defined income-generating activity) is as follows:

$$TV = \frac{[\text{NPV of income-generating activity}]}{[\text{PV of \$1 growing at the same rate as environmental benefits}]}$$

The use of TVs inevitably requires a subjective assessment of what is large or small. Nevertheless, in some instances such subjectivity can be guided by readily observable benchmarks such as house prices and average incomes. Should the TV turn out to be trivially small, then decision-making is simplified. Besides, should the income-generating activity have alternatives of its own—for example a windmill as an alternative to a coal power station—then the TV can reduce even further.

Box 7.2 A threshold value for the old growth forests

Let us return to the example in box 7.1 dealing with wood-chipping in Eden. It was indicated that the present value of the OC of preservation was $15.67 million, if an alternative logging option were available and we had assumed a discount rate of 7 per cent. Assuming this same discount rate: the present value of $1 growing at a very small rate per year, say 0.0005%, over a 17-year period is $10.447. In this context the threshold value of preservation is:

($15.67 million)/10.447　=　$1.5 million

That is, if it had been possible to demonstrate that the benefits of preservation in 1987 were in excess of $1.5 million, then preservation would have taken precedence over logging.

The definition of an OC can be articulated to measure the value of the risk of an environmental accident. Consider the case of business using a natural endowment such as a river or a beach for its day-to-day operations; for example a winery uses a river, or a hotel uses a beachfront. If the business does not take adequate precautions to safeguard the natural endowment, then it will suffer business losses. For example mining operations by BHP in Papua New Guinea and Esmeralda in Hungary resulted in the leakage of cyanide into river systems, causing incalculable environmental damage. In such cases it is possible to value the precaution that needs to be taken against an accident as a threshold value.

Let P represent the probability that the natural endowment would fail in its current management context. Then $(1-P)$ represents the probability that the endowment would not fail. Also suppose that V_D is the present value of the income generated by the business over a specific period of time (say 10 to 15 years), and let K_D be the present value of an income loss should the endowment fail in the present time period. Hence the expected value of income $[E(V_I)]$ for the firm is given by:

$$E(V_I) = \{(1-P) * V_D\} + \{P * (-K_D)\} = V_D - \{P * (V_D + K_D)\} \qquad (7.9)$$

Suppose now that an International Standards Office (ISO) certification requires that the probability of failure be reduced from P to P_S. In order to achieve this reduction, the manager has to sacrifice a certain amount of income, say \$X. The revised expected value of income $[E(V_{IS})]$ would be:

$$E(V_{IS}^*) = \{(1-P_S) * (V_D - X\} + \{P_S * (-K_D)\} = V_D - X - P_S V_D + P_S X - P_S K_D \quad (7.10)$$

It is possible to estimate a threshold value for X on the premise that the revised expected value should exceed the initial expected value. That is:

$$(V_D - X - P_S V_D + P_S X - P_S K_D) \geq V_D - \{P * (V_D + K_D)\} \qquad (7.11)$$

Hence it follows that:

$$X \geq \{(V_D + K_D) * (P_S - P)\} / (P_S - 1) \qquad (7.12)$$

Thus it is possible for the manager to know the size of the sacrifice that has to be made when information on the probabilities and incomes and losses is known.

Concluding remarks

The review of the various methods of valuation presented above reveals that almost all methods contain varying degrees of deficiency. For example, while CVM can theoretically deal with all attributes of the environment, the wide variations in expressed values and the inconsistencies between hypothetical transactions and economic commitments renders the method inefficient. Similarly, the game theory method, apart from being time-consuming, is also limited by the hypothetical nature of transactions. As noted earlier, the travel-cost method does not adequately deal with all features of the environment and cannot be applied unless the environment offers recreational facilities. The hedonic price method too is limited by spatial restrictions and the influences of other socioeconomic variables. The success of methods involving production functions and isoquants will depend on the availability of data on EQ, at least on a cross-sectional basis. At the other end of the spectrum, the methods based on opportunity costs will always tend to provide underestimates of the value of environmental goods and services.

There is an emerging tendency among decision-makers to merely seek a number for the value of environmental outcomes in order to portray the image of having demonstrated environmental awareness. Further, there is also the danger of controversies due to conflicting values presented by different interest groups, especially in the context of sensitive policy decisions; for example the determination of compensation payments by Exxon-Valdez for the Alaskan oil spill. Therefore there is a clear need to move towards less contentious methods of valuation. The use of threshold values in conjunction with other methods might reduce the extent of difficulties.

We now turn to the formulation of specific policy frameworks. In many instances, the application of these frameworks to determine the policies for implementation will rely on the methods of valuation that can be applied to a given context.

REVIEW QUESTIONS

1 An environmental economist collected the following data from seven samples of respondents in applying the binomial approach to Contingent Valuation in order to estimate the monetary value of an environmental endowment. The size of each sample was the same, namely thirty respondents.

Sample number	WTP value	Number of persons willing to pay
1	$1	30
2	$5	27
3	$10	24
4	$15	21
5	$20	15
6	$25	9
7	$30	3

Illustrate how the above data could be used in estimating the value of the endowment. Also indicate whether the binomial approach to contingent valuation could overcome the problems of 'anchoring' and 'embedding'.

2 A popular beach area is also the site for a sewage treatment plant's (STP) ocean outfall. A decline in the amenity value of the beach area has been attributed to the age and outdated technology of the STP. Suppose that the upgrading of the STP and its infrastructure will result in the appreciation of property values in the vicinity of the beach area. Illustrate how the hedonic pricing method could be used to value the environmental benefits of upgrading the STP.

3 Provide three examples where the OC and TV principles could be applied to resolve environmental decision-making problems.

8

Policy Frameworks—Resource Management

We will continue to use figure 2.3 as the basis for our illustrations. Recall that in chapter 2 arrows RA and W were regarded as a composite service flow from the environment to the economy. The aim of this chapter is to illustrate how the market model and its adaptations to the context of public goods and externalities can be extended to the case of specific environmental goods. These are goods that have been traditionally classified as non-renewable and renewable resources. As the term implies, non-renewable resources are finite and therefore can be exhausted. Renewable resources possess the property of regeneration, and thus with careful management could be sustained over time.

Objectives of resource management

Although it is easy to distinguish between renewable and non-renewable resources, two important points need to be acknowledged.
1   The dividing line between a resource being renewable or non-renewable is necessarily fine. This is because the overutilisation of a renewable resource could easily transform it into a non-renewable one. The Blue Whale and certain species of fish are classic examples of this.
2   The utilisation of non-renewable resources without adequate safeguards can also transform renewable resources into non-renewable ones. This is due to the range of harmful environmental externalities that are associated with the use of non-renewable resources. For example the use of coal for power generation without proper filtration devices can permanently impair the capacity of an air-shed to provide clean air.

Because of issues like these, the management of natural resources is not confined to the objective of economic growth (EG) alone. It also includes the objectives of environmental quality (EQ) and intergenerational concern (IGC).

The EG objective is concerned with achieving increases in real national income. Should we be preoccupied with this objective alone, then we are likely to exhaust our stock of natural resources. The singular pursuit of the EG objective can readily transform renewable resources into non-renewable ones and result in the premature depletion of all resources. Unfortunately, this has been the case over most of recent history. When industries and firms, and even households, have had to deal with environmental goods, they have either taken them for granted or exercised stringent cost-cutting measures. For example the unconstrained dumping of untreated effluent into our oceans and waterways as a cost-cutting measure has seriously affected the renewability of our water resources. Firms have for far too long (at least until the late 1980s) regarded filters on smoke stacks and effective emission technologies as investments that are too costly. As a result, we now have acid rain and an incapacitated air-shed as an inevitable part of our lives. Examples of such management practices with their adverse environmental consequences have been, and still are, far too numerous. It is therefore imperative that the list of management objectives should be expanded to include EQ and IGC.

The importance of the EQ objective becomes evident from our foregoing discussion. The scope of this objective is indeed broad. In many situations we need to restore and enhance the quality of our environmental resources. For example we need to clean up our polluted waterways, and we need to enforce strict emission controls so that our cities can be free of smog. In other situations we may have to preserve our environmental resources. For example the unrestricted harvesting of our forestry and fishery resources can threaten their renewability, and so their preservation becomes important. Further, we cannot deal with the EQ objective in terms of individual resources, but must deal with it as a system of resources. Although we deal with renewable and non-renewable resources under separate headings in this chapter, it is important to note that resources constitute complex interactive systems. For example a forest is not merely a collection of trees that make up a renewable resource stock; it is also a complex ecological system of trees of various species, soil and other endowments. Hence, the scope of the EQ objective is to restore, enhance and preserve the quality of our natural resource and ecological systems.

As the term implies, the IGC objective arises from possible conflicts between generations in terms of resource requirements. This conflict has been perceived especially in the context of energy resources. There has been, and still is, a concern that most of our present non-renewable energy resources will eventually be exhausted. For example consider what Lecomber (1979), a British economist, had to say more than twenty years ago: 'A millionfold expansion of total non-renewable energy extends resource life by rather under 200 years. It is difficult to escape the conclusion that before long, either radically different sources of energy or a slow-down of growth is inevitable.'

Even if energy supplies as a whole may appear to be adequate for another two hundred years, the IGC objective is particularly relevant with a resource such as crude oil. Predictions concerning future oil supplies vary, but the dominant contention is that existing oil reserves will be totally depleted by the latter half of the twenty-first century. However, the dominant issue today is not so much the finiteness of the energy resources with proven technologies, as the environmental externalities such as the greenhouse gases that stem from their use and will adversely impact future generations.

The relevance of the IGC objective is of course not confined to energy resources. Deforestation is indeed a global problem. Nadkarni (1987) has estimated that in India some five million hectares of forest were cleared between 1950 and 1970. In Indonesia and Malaysia, the rate of rainforest clearing has been estimated to be 7.5 square kilometres per day (Anderson & Thampapillai, 1990). As Hecht (1985) has pointed out, deforestation has indeed occurred at a massive rate in Brazil—nearly a million hectares a year in the eastern Amazon basin. Such intensive rates of deforestation do indeed threaten future supplies of timber. In the West African country of Malawi the acute shortage of timber already began in the 1980s (French 1986). There are of course numerous other examples. The need for international conferences and agreements on whaling has been prompted by the need to prevent the extinction of whales. The present focus on soil erosion in our agricultural lands stems from the awareness that we may not be able to sustain our agricultural output unless we can conserve the fertility and quality of our soils. There are serious concerns among agriculturists that our traditionally productive areas, such as the Darling Downs and the Murrumbidgee Irrigation Area, need the adoption of urgent conservation measures if they are to remain viable farming areas in the future. To summarise, the IGC objective strives to ensure that the availability and productivity of our resources is sustained between generations.

We can see that there is a substantial amount of complementarity between the objectives of EQ and IGC. That is, preservation of our environmental resources also ensures that these resources are made available to future generations. However, these two objectives together are often in conflict with the EG objective. In the next section we shall consider frameworks that deal with the relationships between the objectives of EG and EQ. After that we shall introduce some frameworks that also include the IGC objective.

Conflicts between EG and EQ objectives

There are two types of frameworks that enable the analysis of conflicts between economic growth (EG) and environmental quality (EQ). The first of these is based on the simple model of market equilibrium. The next is a related framework, which enables the analysis of conflicts that occur between groups of people due to the recognition of the EQ objective. In order to facilitate exposition, we shall look at these frameworks in the context of an example.

The framework based on the market model

Suppose that a factory is located on a river bank upstream. This factory produces a consumer good that we shall refer to as M. The manufacture of M also results in a toxic by-product, which the factory conveniently dumps into the river. Now suppose that the factory has been told to install a water treatment plant upstream to clean the water that flows downstream, because there is a recreational area further down the river. Also assume that the price of M is $20 per unit.

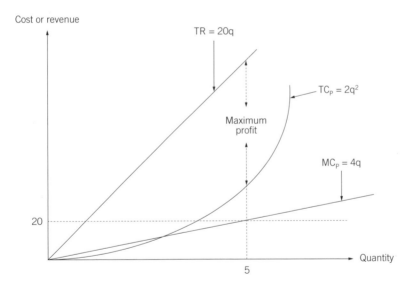

Figure 8.1 An illustration of profit maximisation (without environmental costs)

Consider first figure 8.1, which represents the costs and revenue that are associated with the production of M without any reference to environmental costs. Given that the price of M has been already determined, the total revenue (TR) curve is a straight line, and its equation is:

$$TR \; = \; 20q \tag{8.1}$$

where q represents the quantity of M produced. Let us suppose that the total cost of producing M (without environmental costs), TC_p, is defined by the equation:

$$TC_p \; = \; 2q^2 \tag{8.2}$$

The marginal cost of producing M, MC_p, can be now defined as the first derivative of the equation for TC_p. That is:

$$MC_p \; = \; dTC_p/dq \; = \; 4q \tag{8.3}$$

Note that in figure 8.1 profit is maximised at the point where the distance between the TR and TC_p curves is at a maximum; that is, when the production of M is equal to 5 units. Also note that the point of maximum profit is the point at which the horizontal line representing the price of M intersects the marginal cost curve.

Now suppose that the objectives surrounding the production of M are expanded to include EQ. That is, the factory is told to treat the water that gets contaminated. Suppose that the total cost of water treatment (TC_w) is defined in terms of the quantity of M produced as:

$$TC_w \; = \; q^2 \tag{8.4}$$

Then the marginal cost of water treatment, (MC_w), would be:

$$MC_w \; = \; dTC_w/dq \; = \; 2q \tag{8.5}$$

Given that the management objectives now also include EQ, we need to define the total cost of producing M as:

$$TC \; = \; TC_p+TC_w \; = \; 2q^2+q^2 \; = \; 3q^2 \tag{8.6}$$

The marginal cost of producing M in this context is:

$$MC \; = \; dTC/dq \; = \; 6q \tag{8.7}$$

We can clearly see in figure 8.2 that maximum profit has fallen relative to the situation when treatment costs were ignored, and that the owners of the factory have to cut back production. That is, by incorporating the EQ objective by way of the treatment costs, the production of M falls to 3.3 units.

However, analysis of the type that was shown above presents a problem, in that the owners of the factory will not usually include EQ in their list of objectives unless they are compelled to do so. Further, making the owners of the factory bear the treatment costs also implies that those who own the factory do not have any ownership rights of the river whatsoever. But then, if the downstream users of the river also have no ownership rights over the river, then the

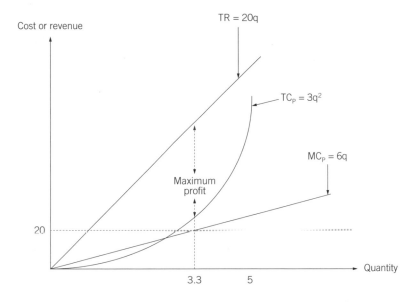

Figure 8.2 An illustration of profit maximisation (with environmental costs)

owners of the factory can demand that the downstream users must also contribute to the treatment costs. The factory might then attempt to raise the price of M. This would be possible if the factory were the only producer of M or if there were a limited number of producers. What we can infer from the foregoing is that the recognition of the EQ objective will almost invariably result in conflicts between different groups. In our example, the conflict would be between the owners of the factory and those who use the river downstream for recreation. The next framework that we consider is one that attempts to address these conflicts between groups.

The framework for conflicts between groups

We shall continue to use the example of the factory upstream and the recreational area downstream. The source of the conflict is primarily the question: Who should pay for the treatment of water that is contaminated? The framework for the analysis of this conflict was first illustrated by an economist named Coase (1960). Now examine figure 8.3 (p. 98). This figure shows two curves, namely the marginal cost of controlling water pollution (MCP), and the marginal benefits of pollution control (MBP). The MCP curve is the marginal cost of water treatment, but this time it is being defined in terms of the quantity of water that is treated. So the MCP curve defines the costs that will be incurred by the factory to control the pollution.

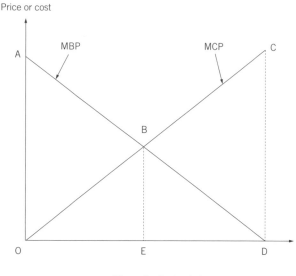

Figure 8.3   An illustration of the Coase theorem

The MBP curve describes the marginal benefits that accrue to those who partic- ipate in recreation; it can be derived by applying some of the methods that were described in the previous chapter. For example suppose that we applied the travel- cost method to the downstream recreational site and derived a demand curve for recreation. Recall that the demand curve for recreation was defined in chapter 7 as a relationship between the cost of travel (which is a proxy for WTP for recreation) and the number of visits (which is a proxy for the quantity of recreation). If we assume a 'one to one' relationship between the number of visits and the quantity of water of water to be treated, then we can assume the MBP curve to be the demand curve for recreation measured in terms of the quantity of water that is treated. Even if such an assumption were unreasonable, it is possible to ascertain a relationship between the possible number of visits and the amount of water to be treated. The underlying premise is that, when the quantity of water treated is small, the facilities for recre- ation will be scarce, and hence the WTP for recreation will be high. On the other hand, when a relatively larger amount of water has been treated, the facilities for recreation are not so scarce, and therefore the WTP for recreation will be relatively low. So this premise is consistent with the law of diminishing marginal utility.

Consider first a situation in which the river is fully owned by the factory. In this case, the factory will refuse to implement pollution control unless it is paid for effecting such control. So those who engage in downstream recreation will have to pay the factory the amount of money that is required to achieve the desired level of water treatment. If the downstream users want OD litres of water treatment, then they will have to pay the factory an amount that is equal to the area below the MCP curve up to OD litres, namely area OCD. Remember that the area below a marginal cost curve measures the total cost, and the area below the marginal benefit curve measures the total benefit. Hence, OD units of pollution control will give the downstream users a total benefit that is equal to the area below the MBP curve, namely area OAD. If the downstream users are prepared to put up with a level of pollution control that is less than treating OD litres of water, then they will incur two types of costs, namely:

- costs in terms of recreational benefits that are sacrificed
- the pollution control costs that have to be paid to the factory.

For example, if the downstream users are prepared to live with ED litres of contaminated water (that is, if only OE litres of water are treated), then their costs are:

- area BED in terms of recreation benefits that they have to sacrifice
- area OBE in terms of treatment costs.

Given that the downstream users will incur these two types of costs, it also becomes evident that the maximum cost saving to these users occurs when they settle for pollution control that is equal to treating OE litres of water. This maximum cost saving is represented by area BCD, and is defined as follows:

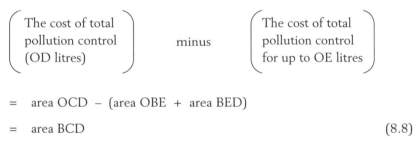

$$= \text{area OCD} - (\text{area OBE} + \text{area BED})$$

$$= \text{area BCD} \tag{8.8}$$

Should the downstream users choose to pay for any other level of pollution control, their cost savings will be smaller. An alternative solution is that the downstream users settle for paying for OE litres of treatment, as it yields them maximum net benefit. The net benefits to those engaged in recreation could be defined as:

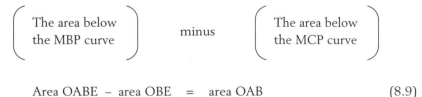

$$\text{Area OABE} - \text{area OBE} = \text{area OAB} \tag{8.9}$$

Note that area OABE represents the total benefits that downstream users get from OE litres of treatment, and area OBE is the total cost of treating OE litres, which cost they have to pay the factory because the factory owns the river. Their payments for any amount of treatment that is either larger or smaller than OE litres will result in a net benefit smaller than area OAB.

Now consider a situation where the river is fully owned by the downstream users. In this case, if the factory wants to pollute the river, then it will have to pay the downstream users an amount of money that represents the loss of recreation benefits. This is of course possible only if the downstream users are willing to accept such payments. Should this be the case, the largest compensation that the factory can pay is defined by the area OAD, and this will happen if the factory decides to produce units of M that correspond to the contamination of OD litres of water. However, the factory can achieve some savings in the size of this compensation cost, if it decides to incur some pollution control costs. For example when pollution control amounts to treating OE litres of water, the costs incurred by the factory are:

- area OBE for treating OE litres of water
- area BED as compensation to the downstream users for the recreation benefits they have to forgo.

The factory will experience maximum cost savings if it treats OE litres of water and compensates the downstream users for the contamination of ED litres of water. This maximum cost saving can be defined as:

$$= \text{Area OAD} - (\text{Area OBE} + \text{Area BED}) = \text{Area OAB} \qquad (8.10)$$

The factory will experience a cost saving that is smaller than area OAB, should the level of pollution control be anything other than OE litres.

What we have seen from the above discussion is that the OE litres of water treatment is the optimal amount of pollution control, regardless of who has the ownership rights to the river. However, in practice there are several impediments to achieving the optimal level of pollution control. To begin with, it is difficult to bestow the ownership of an endowment such as a river to a group of individuals or a firm. This problem is compounded when there are several factories along the banks of the river, and several groups of other users (such as recreationists, anglers, and marine biologists). Further, even if the ownership rights were vested with one group among those involved in a conflict, it is not certain that an optimal solution may be reached by negotiations involving compensation. This is because of the time lags that usually occur with compensations, such as disability payments and accident compensations. However, regardless of these difficulties the analysis of the framework provides some useful basis for policy. If we are able to construct the MCP and MBP curves, then we can use area OBE as the basis for a tax on pollution, or we can use OE litres of treatment as a standard for legislation. The derivation of these taxes and standards would of course become difficult if there were too many polluters and too many affected parties.

Conflicts due to intergenerational concern

As indicated, the intergenerational concern (IGC) objective pervades issues that address the question of resource management over time. This objective is pertinent in the context of managing non-renewable as well as renewable resources. For example the South Pacific Forum has repeatedly expressed an opposition to drift-net fishing because this method overexploits fishery resources, resulting in the possible decline of supply in the future. It is on similar grounds that many conservationists oppose clear felling in forest management. The application of the IGC objective to the management of renewable resources is often directed at achieving *sustainable production*. With non-renewable resources, as indicated earlier, the concern is that we could exhaust our supplies, and that future generations may be unable to meet their own needs.

The economist's approach to the management of resources over time, regardless of whether they are renewable or non-renewable, involves the application of what is called *optimal control theory*. We shall illustrate how an important concept that is employed in this theory can be used in the adaptation of the basic market model that we are familiar with. The main concept that warrants familiarity, for our present purposes, is that of the *marginal user cost*.

A framework for non-renewable resources

So far, in the market models that we have considered, we have not dealt with the issue of time. We have in fact assumed that time was not an issue. We shall now introduce time into our framework, but for reasons of simplicity we shall assume that there are only two time periods, namely the present and the future.

Figure 8.4a shows the demand and supply curves for a given resource during the present time period. Figure 8.4b shows the market demand and supply curves for the same resource in the future period. We can suppose that these future demand and supply curves have been derived by the application of some forecasting methods. Note that the quantity OS, which is shown on the horizontal axis of both figures 8.4a and 8.4b, defines the finite stock of the resource. However, if we have used up some of the resource stock in the present period, then the stock level that is available for the future period will be lower than OS. Alternatively, if we had not tapped the resource during the present, then the maximum possible stock level for the future is OS. From figure 8.4b, we can infer that the future generation requires OQ_1 units of the resource, and from figure 8.4a, we can infer that the present generation requires OQ_0 units of the resource. The conflict between generations would be absent if the sum of the resource requirements, namely $(OQ_0 + OQ_1)$, was less than OS, which is the size of the finite stock. The perception, with respect to some of our essential resources such as crude oil and the ozone layer, is that the stock level falls short of the sum of the resource requirements. Now consider the quantity Q_fS in figure 8.4a and suppose that this quantity is equal to OQ_1 in figure 8.4a. That is, Q_fS represents the quantity of the resource that is required by the future generation. So if resource extraction during the present period exceeds OQ_f, then we will be performing this extraction at the expense of the requirements of the future generation. For this reason, we have an upward-sloping curve that starts at Q_f. This curve is labelled MUC (marginal user cost).

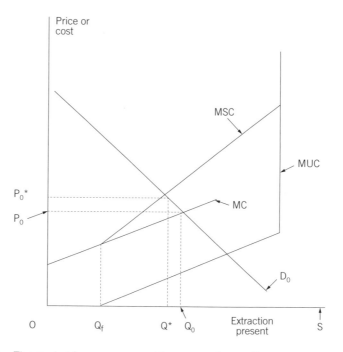

Figure 8.4A   The market for a non-renewable resource (present)

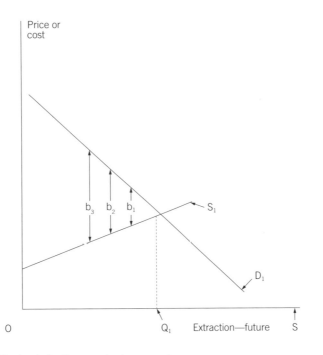

Figure 8.4B The basis for the marginal user cost

The MUC defines the net benefits that future generations must sacrifice for every unit that is extracted during the present period.

The MUC curve is in fact derived from the demand and supply curves that are shown in figure 8.4b. For purely illustrative purposes only, we have shown three vertical lines between the demand and supply curves in figure 8.4b. The following explanation is not precise in a mathematical sense, in that we need to explain the MUC in terms of areas rather than heights. However, for illustrative purposes we shall deviate from precision. When the future requirement, namely $Q_fS = OQ_1$, is left intact, the MUC is zero. When the first unit in excess of OQ_f is extracted in the first period, we can define (somewhat imprecisely) the loss in future net benefits as the vertical distance b_1 between the future demand supply curves. Similarly, we can envisage the heights b_2 and b_3 to be the loss in future net benefits when the second and third units are extracted respectively in excess of OQ_f in the present period. So the MUC curve describes those benefits that are denied to future generations. In a strict mathematical sense, the area below the MUC curve in figure 8.4a is equal to the area that is bound by the demand and supply curves in figure 8.4b. Also note that the MUC curve turns steeply upwards—that is, it tends towards infinity— when the quantity of the present extraction approaches OS. Should the level of the present extraction be OS, then the resource stock would be completely depleted, and since a non-renewable resource cannot be reproduced, the cost that is imposed on the future generation approaches infinity.

Now let us consider only figure 8.4a. If the objective of society were to simply maximise present income and nothing else, then it would extract OQ_0 units of the resource and leave the future generation to contend with $(OS - OQ_0 = Q_0S)$ units. Alternatively, if the IGC objective is also included, then the quantity of extraction would be based on the equilibrium between present demand and the marginal social cost (MSC). In this context:

$$MSC = MC + MUC \tag{8.11}$$

Remember that the supply curve describes the marginal cost of extraction. So until OQ_f units of present extraction the MSC curve is synonymous with the supply curve. When extraction exceeds OQ_f, the MSC curve begins to slope more steeply upwards as the MUC is being added to the MC for every extra unit that is being extracted in excess of OQ_f. Therefore, if the IGC objective is recognised, the optimal extraction will be OQ^* units, resulting in the conservation of $(Q_0 - Q^*)$ units. There would also be a price increase from P_0 to P_0^*.

A framework for renewable resources

The framework that is presented here for a renewable resource is similar to that presented above for a non-renewable resource. This is done for reasons of simplicity, although a robust framework would be one that also defined the rate of regeneration of a renewable resource. The MUC is an important component here as well, and emerges due to the potential for *extinction*. Consider figure 8.5, where the change in the size of the resource population with respect to a given size of the population is described. So if we started with a population size of, say, OB, then due to reproduction the population size can change by a positive amount, namely OY. Biologists often refer to this positive change in population size as yield. Although figure 8.5 is over-simplistic, we can infer that a positive yield is possible so long as the population size is maintained between OA and OC. Hence we can say that a *sustainable yield* is possible if the population size is maintained between OA and OC. However, should the population size fall below OA, then as indicated in figure 8.5, the resource would become extinct. We also need to identify a population size at which the potential for extinction becomes imminent. Let us suppose that this size is defined by the distance OD in figure 8.5. That is, if the population size falls below OD, and yet is still greater than OA, extinction would not occur, but the danger of extinction is real since a slight mismanagement of the resource system could lower population size below OA.

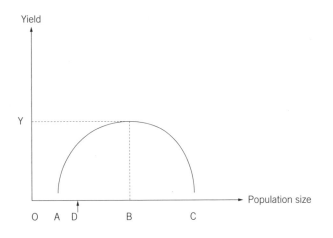

Figure 8.5 Change in size of a renewable resource population

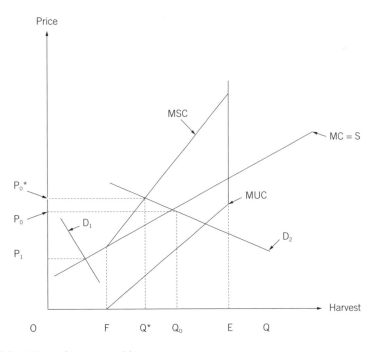

Figure 8.6 Market for a renewable resource

Now look at figure 8.6. The horizontal axis is defined as the population size that can be harvested at a given point in time. Some information from figure 8.5 has been transferred to figure 8.6. Assume that at some point in time the size of the resource stock is OQ and is capable providing sustainable yield. Also assume that:

distance QE in figure 8.6   =   distance OA in figure 8.5

distance QF in figure 8.6   =   distance OD in figure 8.5

Following the explanation that was given for figure 8.5, we can state the following with respect to figure 8.6:

1   When the size of the harvest at the point in time under consideration approaches OF, the potential for extinction emerges. Hence the MUC emerges when the harvest size begins to exceed OF.

2   The MUC approaches infinity as the size of harvest approaches OE because the extinction of the resource becomes a certainty.

So in figure 8.6, the MSC is defined as the sum of the marginal harvest cost (MC) and the marginal user cost (MUC) curves. In this context, the MC curve describes the costs of harvesting the resource. Should the demand for the resource be D_1, then the conflicts between generations would be absent, as harvest size retains the population at a sustainable level. However, conflicts appear when the demand such as that defined as D_2 in figure 8.6. In such a situation, the recognition of the IGC objective enables a reduction in the danger of extinction by prompting a reduction in harvest size from OQ_0 to OQ^*.

Some resource management policies

So far, we have considered some simple frameworks that would enable the analysis of extraction and harvest decisions of natural resources over time. Although more complex frameworks can incorporate the influence of time more explicitly, for our purposes the treatment given here is sufficient. It is also useful to note that the MSC would become larger if we also included a marginal environmental cost or a marginal pollution control cost. This would be the case when the list of objectives involves EG, EQ, and IGC. That is, the inclusion of the EQ objective in addition to the IGC objective would prompt further conservation of our resources. Also, the conflicts between the objectives are invariably influenced by technological change. Therefore technology has an important policy role in resource management.

The frameworks that we examined on pages 101–6 addressed the issue of resource depletion and extinction. In these frameworks we recognised that the costs of extinction or depletion are more likely to be borne by those of the future generation than those of the present. However, our perception of the problems concerning depletion and exhaustion can be altered if we recognise technological change. Some argue that the problem of resource scarcity due to rapid extraction or harvest is not exactly a problem because technology can reverse the scarcity. For example see figure 8.6. Recall that the conflict between generations emerges when present extraction exceeds OF. Now suppose that technological change has resulted in a very cheap substitute for the resource, and as a result the demand for the resource falls from D_2 to D_1. We can now clearly see that technology has eliminated the conflicts between generations and of course the resource price has fallen from P_0^* to P_1.

Renewable resources

Technology can also delay the emergence of potential extinction of renewable resources. Remember that the potential for extinction emerges when harvest size approaches OF, and that extinction becomes certain when the harvest size approaches OE (figure 8.6). Suppose now that, due to advances in genetic engineering, the points of potential extinction and certain extinction shift to the right. In other words, after the appearance of the appropriate technology, a smaller population size is capable of maintaining a sustainable yield. Hence, the harvest quantities for which extinction becomes imminent and certain are larger than OF and OE respectively. The development of aquaculture has been one of the measures that have been widely adopted to offset the potential threat of extinction of renewable fishery resources. However, aquaculture is not without difficulties. For example as Bell (1992) illustrates, accidental leakages from aquaculture sites can significantly contaminate land and water resource systems.

Professor Tor Hundloe of the University of Queensland provides a comprehensive summary of policy measures that could be applied to the context of fishery resources (Environment Australia 1997). These measures, which could be extended to the context of other renewable resources, include: *input controls*, *property rights*, *taxation*, and *output controls*. In figure 8.6, Q^* is the permissible size of the harvest of a renewable resource. As Professor Hundloe argues, it is the determination of this quantity (Q^*) that represents the greatest difficulty. If Q^* has been estimated, then the various measures taken in an attempt to ensure that the harvest size does not exceed this quantity are as follows:

- Input controls deal with restricting the number of persons (and other resources) engaged in harvesting so that the size of the harvest does not exceed Q^*. Usually this is implemented through the issue of a limited number of licences.
- Property rights deal with the rights of ownership or management, endowed to persons either individually or cooperatively, with the expectation that careful rules of management will be developed and exercised. Of course the difficulties that were raised in chapter 3 would remain.
- Tax is the extra levy on resource use to restrict its harvest to Q^*. If we regard the MUC as the extra levy that has to be imposed on the harvesting of a resource, then the tax on a renewable resource will be equal to the price differential $(P_0^* - P_0)$ in figure 8.6.

- Output controls are strict quotas that are issued to the harvesters to ensure that the quantity removed does not exceed Q*. Sometimes output controls are implemented in conjunction with input controls.

Each of the measures contains different degrees of operational difficulties such as unfair distributional effects between groups of people, and the ability to monitor and enforce the measures.

Non-renewable resources

The policy measures for non-renewable resources will be similar to those of renewable resources. However, some important and distinguishing issues need to be considered. With several non-renewable energy resources such as coal and other fossil fuels, the concept of user costs extends beyond the mere depletion effect on the resource. This is because the externalities, which are associated with the extraction and use of non-renewable resources, are capable of reducing the net benefits to future generations and hence are components of user costs. For example the continued use of fossil fuels for power generation and transport has resulted in the build-up of greenhouse gases and associated problems such as the hole in the ozone layer. Hence the policy directive with non-renewable energy resources is now increasingly turning to measures such as:
- clean energy technologies; for example reducing the sulphur content of coal
- alternative energy technologies involving wind power and biomass.

Some argue that these alternative energy technologies remain too expensive compared to traditional energy resources such as coal, and therefore these traditional sources will continued to be used. But if all externalities of these traditional sources are properly internalised, then the price differentials may not be so significant. For example the combustion of ethanol produces significantly less carbon monoxide than petrol.

It is also important to note that one important policy measure with reference to non-renewable resources pertains to modesty in the consumption behaviour of societies. In fact, this consideration extends to renewable resources as well. For example we could achieve significant resource conservation and considerably reduce user costs if we substitute public transport for private transport. How does this represent modesty in consumption behaviour? While trains and buses run half empty, our roads are choked with cars, each of which rarely contains more than one person. However, the shift to public transport may require some incentives such as cheaper prices for public transport and steeper prices for private transport. A good example of a successful public transport system can be found in the island economy of Singapore, which operates a network of buses and Mass Rapid Transit (MRT) rail transport. The maximum price on this transport network for a single trip does not exceed two Singapore dollars. Some market analysts may argue that such price ceilings may prompt market inefficiencies. However, if one argues that such ceilings are the result of internalising the externalities associated with private transport, then they are indeed efficient social policy measures.

To conclude, the formulation of policies for natural resource management needs to take into account all three objectives (EG, EQ, and IGC) concurrently. The measures inevitably have to involve a mixture of proper pricing, regulation, innovation, and attitudinal changes. We examine some of these policy measures in more detail in the next chapter. See the report in box 8.1, which illustrates this mix.

Box 8.1 Israel sets the standard in water recycling

This example is an excerpt taken from the Australian Broadcasting Corporation's *Landline* program.

In the book of Isaiah, the prophet foretold that the desert shall rejoice and blossom as the rose. In the past fifty or so years, the modern state of Israel has gone a long way towards fulfilling that prophecy. But there have been neither deluges of biblical proportions nor more subtle climatic shifts to moisten the scorched earth.

What's changed the face of the desert is an appreciation of the power of such a vital and scarce resource and determination to make every drop count. And that means using things more prodigal nations would simply flush away.

Israel has only limited sources of fresh water for a population which is expected to increase by 40 per cent over the next two decades. The Sea of Galilee and the Jordan River are already stressed after years of drought, and the aquifers that supply the bulk of the drinking water are in an equally delicate state. More people, more pollution, and less water available to feed not only the growing masses but also an essential export industry is a worrying equation. But in Israel the solution may be not only disarmingly simple but also, more importantly, sustainable—use the waste to feed the people. The process of weaning irrigators off fresh water has been one of the remarkable successes of Israeli agriculture over the past few decades. Since 1984 the use of fresh water on farms has halved while the value of production continues to climb. There are a number of factors driving the change: the rapid uptake of technologies such as drip irrigation, the industrialisation of farming whereby orange groves are being pushed aside for greenhouses, and a firm government policy to encourage the reuse of sewage effluent.

It is fair to say Israel and Australia look at water and effluent reuse from entirely different policy directions. Israel puts a far higher value on its raw resource—fresh water—while effluent creates additional usable water for economic growth.

Despite being the so-called driest continent, Australia's fresh water is still relatively cheap and effluent is a problem to be tidied up rather than a productive resource. Perhaps the greatest difference is that in Israel the output of a sewage plant is seen as just part of the overall water resource, a resource that's both productive and safe. In Australia, recycled water is largely an entirely different product from the fresh resource, and that may prove to be the biggest waste issue of all.

Reporter: Steve Letts (source http://www.abc.net.au/landline/stories/s303636.htm)

REVIEW QUESTIONS

1 Consider the case of a factory discharging effluent into a river system upstream. The effluent adversely affects downstream users of the river. An environmental economist has estimated that the marginal benefits (MB) of pollution reduction are given by:

$$MB \ = \ 500 \ - \ 0.5Q$$

where Q represents units of water treated prior to reaching the users downstream.

An environmental engineer has estimated the total costs (TC) of water treatment as:

$$TC \ = \ 0.25Q^2$$

Use the above information to illustrate the Coase theorem, and evaluate its relevance in resolving conflicts arising from environmental externalities.

2 Do you think that the marginal user cost is a relevant concept for the management of Australia's coal and oil resources?

3 Would the relevance of the marginal user cost be diminished with Australia's forestry and fishery resources because they are renewable?

4 Explain how the marginal social cost can be derived when IGC is a management objective for: (a) a non-renewable resource, and (b) a renewable resource.

5 Explain, with examples of renewable as well as non-renewable resources, how technology can delay the emergence of the marginal user cost.

9

Policy Frameworks—
Regulation and Instruments

As we saw in the previous chapter, the externalities arising from the extraction and utilisation of natural resources have emerged as important components of user costs. These externalities are a wide variety of pollution problems involving the contamination of: various water resource systems, air-sheds, and a range of land and subterranean resource systems. In this chapter, we examine the policy measures that will directly target pollution-related issues. However, in several instances pollution issues fall outside the realm of control of individual consumers and firms. Hence the measures that deal with the issue of pollution invariably involve some form of government intervention.

We begin with a simple framework in environmental economics that provides a basis to explain how governments tend to intervene. This intervention takes the form of standards, taxes, and penalties, the setting-up of mechanisms for pollution trading, and other incentives and disincentives. When these forms of intervention are unfeasible, sometimes governments sell the rights to manage environmental goods and services through a system of auctions and bidding. Also, given the climate of increased government regulation on environmental matters, many firms have taken it upon themselves to formulate business strategies that comply with international standards such as ISO 14000 and best practice management.

A simple framework for pollution control

Figure 9.1 shows a simple framework that is widely used in the economics of pollution. Along the horizontal axis, we measure the quantity of permissible pollution; that is, the quantity of pollution permitted to be discharged. The vertical axis measures the unit cost of emissions. The curve labelled MC_A measures the marginal abatement costs. MC_A usually includes the additional costs that firms have to incur (in terms of using pollution control equipment or labour, for example) for reducing emissions. Therefore the MC_A is high when the permissible level of pollution is low, and vice versa. Curve MC_P measures the marginal cost of pollution. This includes items such as adverse health effects and the loss of ecosystem functions and other amenities. As expected, the MC_P curve increases as the level of permissible pollution increases.

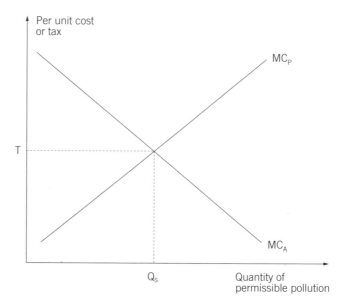

Figure 9.1   Equilibrium between marginal costs of abatement and pollution

When both the MC_A and the MC_P are *measured correctly*, then it is efficient to operate where $MC_A = MC_P$. This is because at this point the total marginal cost of pollution that is $(MC_A + MC_P)$ is at a minimum. Note that we used the term 'measured correctly'. This applies more to the MC_P than the MC_A, because the MC_A is dependent on available technologies and the costs of these are usually known. However, with the MC_P we are likely to make mistakes, because we cannot easily measure all components of this cost. Many of these components, such as ill health and ecological damage, are intangible. Therefore it is more likely that we would end up permitting the emission of more pollution than could be tolerated. That is, if we improve on our information concerning the effects of pollution and our ability to translate these effects into costs, then we will be shifting the MC_P curve to the left, and we will permit a lesser quantity of pollution to be emitted than otherwise. At the same time, if the technologies of abatement continue to improve, the MC_A curve will shift to the left, and we will be able to achieve higher levels of pollution control. What is painfully obvious is that the so-called efficient solution—$MC_P = MC_A$—need not necessarily be a socially optimal one. With improved information and technology we would be approaching this optimum. The equality of MC_P and MC_A merely represents a compromise between the polluters, usually the firms who bear the MC_A, and society at large, which bears the MC_P. This compromise is attained by minimising the total cost. Despite these limitations, the framework in figure 9.1 provides a basis for government intervention and we shall consider this below.

Standards

If measurements have been properly made in order to demonstrate the framework presented in figure 9.1, then the government could either set Q_S as the standard or impose a tax of T to achieve the limit of Q_S. A standard is usually a legal limit on pollution that is set by the government. Any violation of this limit can entail a range of punitive measures, including fines. When there are several polluting firms, the limit of Q_S has to be distributed between the various firms, and the enforcement agency has to monitor the emissions to ensure that the individual limits are not violated. Sometimes standards are used together with taxes and other instruments such as tradeable permits. For example Q_S is the legal limit and a tax of T per unit of pollution is levied for polluting up to that limit, with penalties taking effect when the limit is exceeded.

Environmental quality standards are now strictly enforced through a range of regulatory bodies such as the Environment Protection Authorities (EPA) of various state governments and Environment Australia, which is a Commonwealth agency. Enforcement and regulation are also performed by a range of other bodies, such as catchment management commissions and local governments. Environment Australia has established a National Pollution Inventory (NPI), which is a database of pollutant emissions and discharges in Australia. This database has been used in the setting of a range of environmental quality standards as a National Environment Protection Measure (NEPM). The ambient air quality standards in terms of greenhouse gases and lead are given in table 9.1 (p. 114).

Similarly, water quality standards limit the salt content of river systems to 60 Electrical Conductivity (EC) units. The other pollutants of importance in water quality standards are collectively termed biochemical oxygen demanding (BOD) material. These include: phosphorus, nitrogen, heavy metals, and faecal coliforms.

The enforcement of standards appears to have prompted many firms into self-regulation. One such measure pertains to compliance with a set of standards issued by the International Standards Office (ISO). Governments have also increasingly begun to recognise firms that obtain accreditation from the International Standards

Table 9.1 The NEPM ambient air quality standards

Pollutant	Averaging period	Maximum allowable concentration	Maximum days per year the standard can be exceeded
Carbon monoxide	8 hours	9.0 ppm	1
Nitrogen dioxide	1 hour	0.12 ppm	1
	1 year	0.03 ppm	None
Photochemical oxidants	1 hour	0.10 ppm	1
	4 hours	0.08 ppm	1
Sulphur dioxide	1 hour	0.20 ppm	1
	1 day	0.08 ppm	1
	1 year	0.02 ppm	None
Lead	1 year	0.50 µg/m	None
Particles less than 10 micrometres in diameter	1 day	50 µg/m^3	5

Source: Environment Australia, State of Knowledge Report: Air Toxics
and Indoor Air Quality in Australia, *Canberra, 2001*

Office (ISO) for adopting prescribed enhanced environmental management practices. The argument is that most taxes, standards, and permits achieve only a limited amount of environmental improvement because they target single pollutants or single markets. As a result these approaches are often labelled 'end-of-pipe approach' or 'product approach'. The ISO 14000 series prescribes a 'management systems approach'. For example ISO 14001 specifies the requirements for developing an environmental management system that can be certified by an external party. This includes designing and implementing the business system. When necessary, the ISO 14001 certification also requires the adoption of the following:

i ISO 1421-24 (Environmental Labelling): This defines the criteria for labelling products and defines which products could be labelled, and hence is useful for business managers who are considering product differentiation.

ii ISO 1441-44 (Life-Cycle Assessment): This provides a basis for firms to evaluate the outcomes of their decisions within the framework of the 'cradle to grave' concept; that is, from the use of raw materials to the disposal of the product at the end of its life.

iii ISO 1410-12 (Environmental Auditing): This provides the basis for auditing the environmental management system and the criteria for identifying the persons who could qualify as environmental auditors.

iv ISO 1431 (Environmental Performance Evaluation): This provides guidelines for a business to set up a system to evaluate its own environmental performance through a range of measures and indicators.

The ISO 14001 certification is given for a firm that develops an environmental management system to proactively control factors that would generate damaging environmental effects if they were not controlled. The certification rests on the premise that the firm is able to identify causes of undesirable environmental events and control them early rather than taking action after the event occurs. So the construction of a tailings dam for cyanide treatment in a high rainfall or snow melt area would need to satisfy stringent engineering design specification standards. The management plan must show how it proposes to avoid the collapse of the dam and detail the immediate contingency measures it has in place should a collapse occur. So, in order to obtain the certification, a gold mining firm would have to show how it carried out ISO 1441-44 and at the same time specify how it will put into practice ISO 1410-12 and ISO 1431.

Taxes and charges

Some economists argue that a tax on pollution could achieve a greater level of pollution abatement than a standard would. To illustrate this argument, we need to first consider how a pollution tax is supposed to work.

In figure 9.2, we consider an individual firm and its marginal abatement cost (labelled MC_{A1}). Suppose that this firm's production practices involve maximising profits at an emission level of Q_P units of pollution. Note that at Q_P units of pollution, MC_{A1} is zero, implying that profit maximisation is achieved when no pollution is controlled. If the government now levies a tax of T per unit of pollution emitted, then the business can be expected to respond as follows:

1 Because $(T = MC_{A1})$ at the pollution level Q_{S1} it is cheaper to pay the tax than to abate the pollution up to Q_{S1}, the firm will make a tax payment of (T^*Q_{S1}) for the first Q_{S1} units without abatement.

2 It will clean up the remaining $(Q_P - Q_{S1})$ units of pollution at its abatement costs because it is cheaper to abate this quantity rather to pay taxes for them; that is for $(Q > Q_{S1})$, $(MC_{A1} < T)$

This response of paying taxes for pollution quantity Q_{S1}, and then abating the remaining quantity $(Q_P - Q_{S1})$, is indeed cheaper than paying a tax of (T^*Q_P) for the total quantity (Q_P) of pollution. We can also see how a greater level of

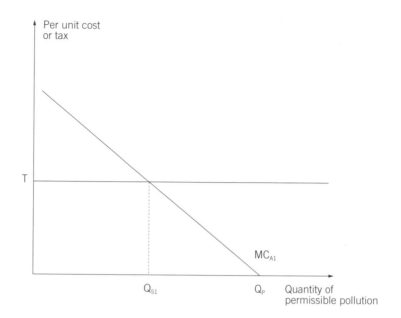

Figure 9.2   Taxes and standards

pollution gets abated with a tax than with a standard that is not accompanied by a tax. Supposing the standard was specified for the firm as Q_{S1}; then the firm can freely pollute up to Q_{S1}, and adopt abatement only when its pollution exceeds Q_{S1}. However, with a tax, the firm will abate $(Q_P - Q_{S1})$ units and the government can use its tax collection to abate further units of pollution.

Some of the environmental taxes that currently operate in Australia are:

- effluent charges
- emission taxes
- load-based licensing schemes covering air and water pollution
- product tax on ozone-depleting substances.

Effluent charges, emission taxes, and load-based licences can be easily related to the framework in figure 9.2. That is, the polluter's liability is determined directly in terms of the quantity of pollution and the predetermined tax rate. For emission taxes to work, it should be possible to measure the levels of emissions such as discharges into air-sheds and water and land resources. Usually the emission taxes are applied in the context of points and stationary sources of pollution, due to the ease of monitoring. With non-stationary sources of pollution such as moving vehicles, it is easier to apply a product tax, so that liability is transferred to the price of the product. In the case of product taxes, those commodities that cause more harm to the natural environment would command a higher price. For example in Australia (as in most other developed countries) the price of leaded petrol is higher than that of unleaded petrol, due to a product tax.

The potential to achieve greater efficiency through taxes on pollution depends very much on how these tax earnings are used in the wider community and on how firms respond to the taxes. For example Hamilton, Hundloe, and Quiggin (1997) argue that environmental taxes, if used sensibly, can benefit society in a wide range of ways, including an increase in employment and the promotion of cleaner production methods. In a modelling exercise Hamilton, Hundloe, and Quiggin (1997) illustrate that if environmental taxes are returned to the community and labour and payroll taxes are removed, then a net gain in employment that ranges between 100,000 to 150,000 can be achieved. In the context of being faced with taxes and penalties, many firms have responded by improving their production methods. For instance, BHP at Port Kembla has begun reusing waste water for cooling its slag heaps (James 1997). Box 9.1 gives an example of some Swedish firms that have even included salary bonuses for workers who are able to achieve lower levels of emissions.

Taxes, like standards, suffer from operational difficulties. Both require advanced methods of monitoring, measurement, and enforcement. Further, there needs to be in existence a well-established institutional setting. In Australia this role is filled by agencies such as Environment Australia and the Environment Protection Agencies in various states.

Box 9.1 Emissions charges with a twist in Sweden

Combustion plants produce energy—electricity and heat—by burning different kinds of fuel. But during combustion, air-polluting compounds such as nitrogen oxide (NO) and nitrogen dioxide (NO_2), collectively termed nitrogen oxides (NO_x), and sulphur dioxide (SO_2) are released. Since January 1, 1992, large combustion plants have paid an environmental charge on NO_x emissions. 'Large' plants are defined as having a capacity of 10 MW or more and an annual energy production exceeding 50 GWh. Smaller combustion plants are not liable because of the higher relative cost of continuously measuring the emissions. The charge SEK 40 (US $4.80) per kilogram of NO_x is not a tax. Instead it is redistributed among liable plants in proportion to their energy production. As a result, plants which produce much energy relative to their total emissions benefit, while those with a low ratio of energy to emissions lose. Some plants earn money from this system while others underwrite it.

Most of the liable combustion plants are found in energy production, that is, heating and power plants. The pulp and paper industry, the chemical industry and the metal industry also have combustion plants for energy production. Waste incineration plants producing energy are similarly liable for the charge. There is a wide variation in net payment (charge minus refund) within the industries. For example energy production plants range from making a net payment of SEK 10m ($1.2m) to receiving a net income of SEK 14m ($1.7m). In 1992, approximately SEK 100m ($12m) was redistributed. The refund system was necessary in order to achieve a fair system. The competition between small (non-liable) and large (liable) combustion plants would have been distorted if the charge was not refunded to the liable plants. The fact that the charge is refunded and thereby only has an environmental purpose has facilitated acceptance of the charge. A positive side effect is that less polluting plants are favoured economically and thus given a competitive advantage. The refund system has contributed to the considerable success of the charge. Many companies started NO_x-reducing projects as soon as a parliamentary decision was taken in 1990, in order to have as low emissions as possible when the charge came into force in January, 1992. The management and the operators at the plants have become more focused on reducing NO_x. At one plant the operators are given a salary bonus if NO_x emissions are low. Though the combustion plants are given an economic incentive to reduce their emissions, they are not forced to do so by regulation. It is up to the individual plant to decide. Companies can choose whether to reduce their NO_x emissions or pay the charge. Generally speaking, the liable plants have a greater incentive to seek ways to reduce emissions than any government body.

This example is an excerpt taken from the following web site, which is maintained by the International Institute for Sustainable Development, Winnipeg, Canada: http://iisd1.iisd.ca/greenbud/nitro.htm)

Emissions trading

To illustrate the principles underlying emissions trading, suppose that the industry consists of only two firms. Further, assume that the government's target is to achieve an overall standard of Q_S. If this standard is equally divided between the two firms, then the pollution limit on each business is $(Q_S/2)$ and the government sells to each firm pollution permits to the value of $\{T_0{}^*(Q_S/2)\}$, where T_0 is the per unit value of a pollution permit. As in the previous section, we will also suppose that both firms maximise their profits if they are each allowed to pollute up to Q_P units of pollution, and of course $Q_S <$ {Total Maximum Potential Pollution Load = (2QP)}.

Consider figure 9.3, where MC_{A1} represents the abatement costs for Firm 1, which uses more expensive, older technology than Firm 2, whose abatement costs are MC_{A2}. Suppose that the firms face a per unit pollution charge of T_0. Following the line of reasoning presented above in the section headed 'Taxes and charges', Firm 2 would prefer to discharge Q_{P2} units of pollution with its pollution permits and control $(Q_P - Q_{P2})$ units of pollution using its own technology. Similarly, Firm 1 would prefer to control $(Q_P - Q_{P1})$ units of pollution using its own technology and discharge Q_{P1} units of pollution at a per unit cost of T_0. Given that each firm has pollution permits for controlling $(Q_S/2)$ units of pollution, we find that Firm 2 has a surplus of permits {$(Q_S/2) - Q_{P2}$}, while Firm 1 has shortage of permits {$Q_{P1} - (Q_S/2)$}. As long as {$(Q_{P1}+Q_{P2}) < Q_S$}, Firm 1 can decrease its total costs by buying permits from Firm 2. This in essence is the principle underlying emissions trading. That is, less efficient firms need not necessarily leave the industry in the context of stringent standards. They can prolong their existence as long as there are some efficient firms in the industry, and this extended time period could be used for them to make the necessary adjustments.

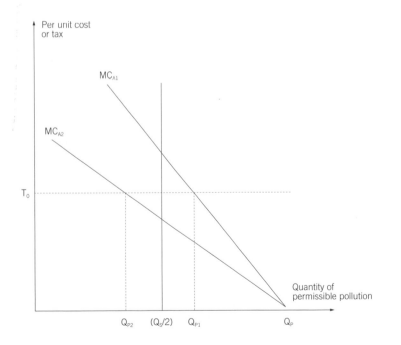

Figure 9.3 The basis for emissions trading

Tradeable permits have been in operation in the United States for nearly two decades and have been implemented by the US Environment Protection Agency under two programs, namely *bubbles* and *offsets*. A bubble refers to a specific geographic region; those who trade within the bubble aim to reduce the pollution levels of the specific region. The offset programs apply to new firms, generating new emissions; these are permitted to operate only as long as they are able to reduce emissions from existing sources. Large firms have also used both the bubble and offset frameworks for internal trading. That is, they trade pollution permits between various branches of the same firm. The significant advantage of emissions trading has been the savings in abatement costs. Hahn and Hester (1989) estimate the cost savings for hydrocarbon emissions in one corporation alone to be around US$60 million. However, despite these stated advantages, there are difficulties in this system that pertain to monitoring, detection, and enforcement. James (1997) describes in detail the tradeable permit schemes that are operational in Australia. Some of these are summarised in box 9.2.

Box 9.2 Some Australian examples of tradeable permits

- **The Hunter River Salinity Trading Scheme:** This scheme involves some 1,000 salt credits, and the standard is set by the Department of Land and Water Conservation (DLWC), which is to keep the salt level in the river with 900 Electrical Conductivity (EC) units. The permit holders are mainly coal miners and power stations. Monitoring has been possible because each credit holder is assigned an authorised point of discharge and monitoring gauges have been set up to automatically report river salt levels to a central data warehouse.

- **The Murray–Darling Basin Salinity Scheme:** Here the tradeable system does not involve exchanges between individuals or industries but between the state governments of South Australia, Victoria and New South Wales. In this scheme, each state earns salt credits by implementing investments and other management activities to reduce salt levels in the river. For example the investments could involve planting eucalyptus trees. Debits accrue when high levels of salt enter the river system via drainage. The administering agency is the Murray–Darling Basin Commission.

- **South Creek Bubble Licence Scheme on the Hawkesbury–Nepean river system:** This scheme involves an imaginary bubble placed around sewerage treatment plants that are located within the Hawkesbury–Nepean river system. The aim of the scheme is to reduce phosphorus and nitrogen levels and the scheme is administered by the Environment Protection Authority of New South Wales. The bubble licence involves the setting up of a discharge limit within the bubble, while the operators can then decide how the limit could be distributed.

- **The Bubble Scheme at Kwinana Industrial Area**: This scheme, which operates in Western Australia to the south of Perth, is similar to the above, but deals with air quality and is hence more complex. The aim is to control the amount of sulphur dioxide emissions in specific areas to an annual limit not exceeding 60 micrograms per cubic metre. The scheme is administered by the Western Australian Department of Environmental Protection. Three specific areas had been chosen for monitoring, namely the area on which the industry is located, a buffer area surrounding the industry, and areas beyond the buffer area.

Source: D. James, Environmental Incentives: Australian Experience with Economic Instruments for Environmental Management, *Environmental Economics Research Paper No. 5, Environment Australia, 1997*

Property rights

Almost all environmental goods and services lack the essential features required for a system of private property rights to function effectively. Recall from chapter 3 that these are *enforceability*, *exclusivity*, and *transferability*. Enforceability means that the rights of ownership can be enforced by law, while exclusivity refers to the exclusiveness of the ownership rights once they have been enforced. Transferability refers to the ability to transfer the ownership rights at a price. These three conditions of property rights are satisfied only when the goods concerned can be exchanged through the market system. That is, a system of private property rights and the market mechanism work together in unison when goods are exchanged in the market.

The property rights approach to pollution problems is based on the premise that if ownership rights to environmental capital (KN) can be offered (at a price) to a private entity, then a market for the utilisation of KN could emerge. The goal of sustaining this market would in turn necessitate the proper maintenance of KN. As indicated in chapter 3, this approach has worked in certain instances such as beaches, lakes, forests, and even river systems and water sheds. The approach has basically involved the auctioning of the item of KN (say a beach or a forest) and offering its management rights to the highest bidder who bids above a reserve price.

But again, as mentioned earlier, there are other difficulties such as the emergence of income inequalities that could be associated with this measure.

Other incentives and disincentives

Governments have also influenced firms and individuals in several other ways. These include income tax credits and subsidies for pollution control, recycling, and other favourable environmental management strategies. Some of the incentive schemes that are currently operational in Australia are:

- rate concessions by local governments for sustainable land management
- subsidies and grants for tree planting and vegetation protection
- environmental performance bonds for firms
- beverage container deposit refund schemes (in South Australia).

While the role of subsidies, concessions, and refunds is self-explanatory, the functioning of performance bonds warrants a brief explanation. The basic principle is similar to that of a rental bond. Firms need to lodge a bond with the appropriate administrative authority, prior to commencing operations, as a guarantee that they will adopt specific environmental protection measures. The bond is forfeited on a proportional basis in relation to the extent of protection measures that have been adopted. The bond is returned in full if the protection measures demonstrate full compliance. In Australia, performance bonds have been used in the context of mining operations. Mining firms have had to demonstrate satisfactory mine rehabilitation measures to recoup their bonds.

In addition to the above, Environment Australia (2001) also reports that the finance sector has begun to include environmental factors in loan and insurance applications. For example firms that demonstrate safeguards against environmental accidents are able to gain lower insurance premiums. One major Australian bank is reported to be offering a cheaper home loan for environment-friendly home designs.

In some instances, governments have been compelled to examine their own actions and demonstrate cross-compliance, especially when some of their actions prompt environmental degradation, while others impose punitive measures for such degradation. For example a farmer may receive a fertiliser subsidy, while at the same time facing a penalty for discharging contaminants into an adjoining river system.

In this and the previous chapter we have reviewed a set of policy measures that could be adopted for managing environmental goods and services. However, we have examined these policies from a microeconomic perspective. Our task now is to examine the development of environmental policies from a macroeconomic perspective. We shall look at this in the next five chapters.

REVIEW QUESTIONS

1 Some authors have argued that the enforcement of standards and strict regulation are good for industry, in that industry becomes more efficient and competitive. Do you agree with this statement?
2 Discuss with examples the advantages and disadvantages of a system of marketable pollution permits.
3 Discuss the feasibility and desirability of imposing higher taxes on private transport and using the tax earnings to subsidise public transport.
4 Illustrate the feasibility and efficiency of using performance bonds on a wider scale than is done in Australia now.

Part 3

Macroeconomics and the Environment

10

Some Important Concepts in Macroeconomics

In this and the chapters that follow, we examine how the environment can be internalised into macroeconomic analyses. We begin with a brief review of some important concepts in macroeconomics. An understanding of these concepts could in turn help us conceptualise the environment in macroeconomic terms.

National product

Macroeconomists regard the entire economy as a *single entity* that produces a *single good*, namely national product. No one has touched or smelt national product. It is the aggregate of all goods and services that are produced in the economy, ranging from the humble tuna sandwich to the wide-bodied aircraft that fly our skies. The price of national product is usually referred to as the price level. General increases in price level reflect a state of inflation in the economy.

The size of national product is usually measured by one of two methods. These are:

- the sum of all real final expenditures
- the sum of all real final incomes.

When an economy is in equilibrium, both methods will yield identical values for the size of national product. It is usual for the sum of all expenditures to be referred to as *aggregate demand* and the sum of all incomes to be referred to as *aggregate supply*.

To illustrate the expenditure method, denote the size of national product as Q_{NP} and suppose that this size is generated by N final expenditures involving the following quantities and real prices of goods $(Q_1, ..., Q_N)$ and $(P_1, ... P_N)$. Then:

$$Q_{NP} = (P_1{}^*Q_1) + \ldots + (P_N{}^*Q_N) \tag{10.1}$$

Similarly, if the production of goods for the N final expenditures depends on K factors, the quantities and real prices (returns or wages) of which are respectively $(q_1, \ldots, q_N)$ and $(W_1, \ldots, W_N)$, then:

$$Q_{NP} = (W_1{}^*q_1) + \ldots + (W_N{}^*q_N) \tag{10.2}$$

Note that in measuring national product, real prices have to be used and applied to the context of final expenditures or transactions. The concepts of real values and final expenditures will now be considered.

Real values and price level indexes

The prices that are observed in usual market exchanges are referred to as *nominal values*, which include the effects of inflation. Real prices, on the other hand, are ones where the effects of inflation have been removed. It is usual to define real prices as follows:

$$\text{Real price} = \frac{\text{nominal price}}{\text{price level index}} \tag{10.3}$$

Price level is usually quantified as an index of expenditures. The general formula for the price level index of a given year is as follows:

$$\text{Price level index} = \frac{\text{observed expenditure in chosen year}}{\text{expenditure at base year prices}} \tag{10.4}$$

Note that using a real price is the same as using the price of the base year for aggregating the transactions of each year. To illustrate, suppose that we are dealing with a period of T years $(1, \ldots, T)$ and N goods $(1, \ldots, N)$ in each year. If we use double subscripts for quantities and prices in equation (10.1) we will get the following data set:

Year 1: $(Q_{11}, P_{11}), (Q_{12}, P_{12}), \ldots, (Q_{1N}, P_{1N})$ $\tag{10.5}$

Year 2: $(Q_{21}, P_{21}), (Q_{22}, P_{22}), \ldots, (Q_{2N}, P_{2N})$

.

.

.

Year T: $(Q_{T1}, P_{T1}), (Q_{T2}, P_{T2}), \ldots, (Q_{TN}, P_{TN})$

If we nominate Year 1 as the base year, then the size of national product for each year at base year (constant) prices will be:

Year 1: Q_{NP1} = $(P_{11}{}^*Q_{11})$ + ... + $(P_{1N}{}^*Q_{1N})$ (10.6a)

Year 2: Q_{NP2} = $(P_{11}{}^*Q_{21})$ + ... + $(P_{1N}{}^*Q_{2N})$ (10.6b)

.

.

.

Year T: Q_{NPT} = $(P_{11}{}^*Q_{T1})$ + ... + $(P_{1N}{}^*Q_{TN})$ (10.6c)

That is, the same set of prices $(P_{11}... P_{1N})$ is used for each year. Hence, in most countries' statistics, one would observe sets of data presented in constant prices as well as at nominal market prices.

Comparisons of national product measured at nominal prices over time would invariably reveal a significant increase in national product. This is the case when prices increase much faster than quantities of output due to inflationary forces. However, such increases are modest when real prices are used. Hence the measurement of national product at real prices gives a proper appreciation of the performance of an economy. For example an examination of Indonesia's national income accounts reveals that the size of national product at nominal prices in 1993 was seventy times greater than it was in 1973. However, at real prices, there is a mere threefold increase over the same time period.

Final expenditures

As indicated, changes in the size of national product are indicators of the performance of an economy. The use of final expenditures guards against an overstatement of performance. Consider the following example. A farmer has produced 200 kilograms of beans. A leading supermarket chain buys these beans from the farmer at $0.50 per kilogram. The same beans are then transported to a supermarket and sold to customers at $0.60 per kilogram. If all expenditures are counted, then we have ${(0.5*200) + (0.6*200)} = $220. However, in doing so we have counted the same beans twice. Therefore, to avoid an overstatement of national product, only the expenditures pertaining to a good or service at the point of the final transaction are counted. With the example of the beans, if the supermarket transaction is deemed to be the final one, the contribution made by 200 kilograms of beans to national product is $120.

Economic growth

Economic growth is a standard measure of performance for any economy. Simply put, it is the increase in the size of real national product. To be more precise, economic growth should be the increase in real per capita national product; that is, real national product per person in the economy. National product is usually measured at three levels. The simplest level of measurement is the sum of all final expenditures (or factor incomes) within the country during a specific time period, which is usually a year. This is referred to as *Gross Domestic Product* (GDP). For convenience, the expenditures that make up GDP are classified as follows:

* Consumption (C)
* Investment (I)
* Government Expenditure (G)
* Exports (X)
* Imports (M).

The statisticians have various rules and guidelines for measuring these expenditures. For I, for example, only new investments that occur during a given year are counted. An item that entered a firm's inventory in the previous year would have been counted in the previous year and not in the following year, even if a sales transaction occurred in the following year. We shall, however, not deal with these rules and guidelines. A clear explanation of these for Australia is given in Jackson (1989). The definition for GDP is:

$$GDP \;=\; C + I + G + X - M \tag{10.7}$$

The second level of measurement for national product, *Gross National Product* (GNP), takes into account the flows of income in and out of a country. This is defined as:

$$GNP \;=\; GDP - R \tag{10.8}$$

where R stands for 'net factor incomes from abroad'. For example foreign nationals who live in Australia remit some of their income earned in Australia to their homes and at the same time the many Australians who live abroad remit to Australia some part of their income. R is the difference between these two remittances. If Australia's GNP is more than its GDP, it means that remittances into Australia by Australians living abroad are more than the outward remittances of expatriates in Australia.

The third level of measurement is *Net National Product* (NNP). This recognises the wear and tear of the investments or capital stocks:

$$NNP \;=\; GNP - K_c \tag{10.9}$$

where K_C represents the depreciation allowance for capital stocks; it is also sometimes referred to as capital consumption.

Hence in gauging economic growth, NNP is a better measure than GNP, which in turn is a better measure than GDP. However, even many economists concede that economic growth should not be the sole criterion to measure an economy's performance. For example when natural disasters strike, there is increased economic growth due to restoration activities. The environmental economist argues that nature is capital and that NNP should be further refined by subtracting from it the wear and tear of natural capital. We shall discuss this in the next chapter.

The reverse of economic growth is a *recession*. Periods of growth and recession are accompanied by changes in the price level. A sustained increase in the price level is *inflation* and the reverse of this is *deflation*. We present next an elementary review of the causes of these economic phenomena (that is, growth, recession, inflation, and deflation). We shall do this by examining the relationship between the size of national product and the price level.

The relationship between national product and price level

An understanding of the relationship between price level (p) and the size of national product (Q_{NP}) is useful for an illustration of some basic concepts in macroeconomics. The nature of this relationship depends on how national product is measured, namely whether it is measured as the *sum of all real final expenditures* or as the *sum of all real final incomes*.

Q_{NP} measured as the sum of real final expenditures

When Q_{NP} is measured as the sum of all real final expenditures, the relationship between the price level and Q_{NP} is inverse; in other words, as the price level increases the size of $\dot{Q}_{NP}$ decreases; that is, the sum of all real final expenditures decreases. A more rigorous reasoning for this inverse relationship is given in the context of the IS–LM model in more advanced texts. An intuitive and simplistic reasoning for this is as follows. When the price of a single good rises, consumers actively search for substitutes. For example if the price of coffee goes up, then consumers may switch to tea or some other beverage. But when the price level increases, we have a situation where all prices are rising simultaneously. In this context there are three possible substitutes for 'total present expenditures' on items that make up national product. These are:
- holding financial assets including money
- postponing expenditures for the future
- making expenditures in other countries.

So if prices as a whole keep rising, people will prefer to hold their money in various financial assets rather than spending it now, or they may even go overseas where goods are cheaper to make purchases. Hence, in figure 10.1, the relationship between Q_{NP} as the sum of all real final expenditures and the price level is a downward-sloping curve. Because the sum of all expenditures is referred to as aggregate demand, the downward-sloping curves are also labelled aggregate demand (AD) curves.

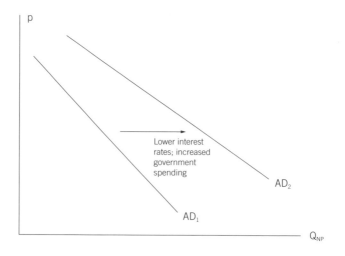

Figure 10.1   Aggregate demand

As shown above, aggregate demand is made up of consumption (C), investment (I), government spending (G), exports (X), and imports (M). We can illustrate the effects of specific macroeconomic policy variables on the components of aggregate demand. For example if income taxes are lowered, consumers have more disposable income and therefore C can increase. Such increases can prompt the AD curve to shift from AD_1 to AD_2, as illustrated in figure 10.1. Similarly, a lowering of interest rates can stimulate an increase in I, while a lowering of exchange rates can encourage an increase in X and a lowering of M. Hence, if policy-makers wish to stimulate economic growth, they can endeavour to do so by stimulating aggregate demand.

Q_{NP} measured as the sum of real final incomes

When Q_{NP} is measured as the sum of all real final incomes, the relationship between the price level and Q_{NP} is positive; in other words, as the price level increases the size of Q_{NP} increases; that is, the sum of all real final incomes increases. Although such a relationship is generally true, this is not the case at all times. The general relationship is believed to hold true in between two extremes, at both of which national product has no bearing on price level. At one extreme (figure 10.2A) the relationship between Q_{NP} and price level is a horizontal straight line. This happens if there is excess capacity in the economy in terms of labour and resources. That is, producers are willing to offer goods and services for sale without any reference to prices. Such a situation normally occurs when the economy is in a severely depressed state.

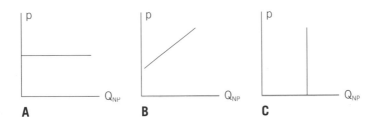

Figure 10.2 The relationship between Q_{NP} and price level. A—with an excess of labour and resources; B—the normal situation; C—with full employment of labour and resources

At the other extreme, the relationship is a vertical line, as shown in figure 10.2C. The situation here is that the economy cannot produce beyond a certain amount of national product because of its resource limitations. This limit is called the *productive capacity* of an economy. That is, regardless of the price level, the economy cannot generate more than a fixed amount of national product. The productive capacity of an economy is usually described in terms of the complete utilisation of labour and capital stock; that is, the full employment of the factors of production. However, the full employment of labour has usually been the more important determinant of productive capacity than capital stock. Figure 10.2B describes the general situation. Here, when price level rises, producers employ more resources to increase output and hence the size of Q_{NP} as the sum of resource incomes increases. Because the sum of all incomes is also referred to as *aggregate supply* (AS), each of the curves in figure 10.2 represents an AS curve. It is customary to describe the AS curve as it appears in figure 10.3, where the features of figures 10.2B and 10.2C have been combined. It is usual to ignore the features of figure 10.2A, because they happen only when economies are severely depressed. The last time such a curve was observed was during the depression of the 1930s. Because Keynes (1936) argued in the context of the 1930s that the AS curve was horizontal, it is also referred to as a *Keynesian AS curve*. The vertical AS curve is referred to as the *classical AS curve*.

In just the same way as taxes and interest rates can prompt shifts in the aggregate demand curve, corporate taxes and wage claims can cause the AS curve to shift.

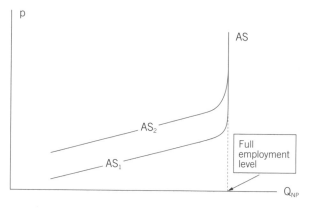

Figure 10.3   Aggregate supply

If workers make large wage claims or if corporate (business) taxes are raised, then the AS curve can shift from AS_1 to AS_2. Note how, in figure 10.3, even if the AS curve has shifted, the productive capacity remains unchanged.

Macroeconomic equilibrium

As in market equilibrium, the point of intersection of the AD and AS curves determines macroeconomic equilibrium (see figure 10.4). At this point:

The size of national product as the sum of all real final expenditures	=	the size of the national product as the sum of all real final incomes

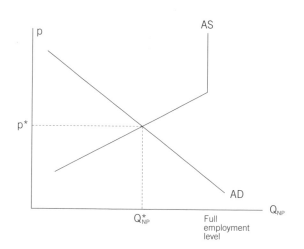

Figure 10.4   Macroeconomic equilibrium

The main purpose of macroeconomic policy is to increase economic growth and employment while also maintaining a state of equilibrium. Being out of equilibrium can prompt a range of difficulties.

> ### Box 10.1 Examples of macroeconomic processes
>
> The framework in figure 10.4 can be used to illustrate some world events. For example in 1973 OPEC, the cartel of oil-producing countries, quadrupled the price of oil. This resulted in an overall increase in all business and production costs. As a result, the AS curve shifted to the left, causing a reduction of national product and an increase in the price level: in short, a recession and inflation. More recently, in South-East Asia, the lack of investor confidence has resulted in the AD shifting to the left, and this has caused a lowering of the price level alongside a recession.
>
> As of July 2000, the Federal Government of Australia introduced a broad-based goods and services tax (GST). The opponents of this tax argued that it would have both a recessionary effect and an inflationary one due to the rise of production costs—that is, a shift of the AS schedule to the left. However, the government was able to partially counter this by:
> - granting some income tax relief
> - increasing spending on the housing sector.
>
> These measures enabled the AD schedule to shift to the right, resulting in a modest amount of growth and an increase in the price level.

Some concluding remarks

We have examined some basic concepts in macroeconomics. It is now possible to illustrate how the conceptualisation of the environment will influence formal macroeconomic analyses. The notion that the environment degrades or depreciates can be incorporated into both AD and AS schedules. At the same time, the argument that we can either scientifically or otherwise restore environmental endowments that were previously not functioning raises the possibility of introducing the concept of 'environmental investments'. Further, recognising that nature is capital can also influence the definition of productive capacity. That is, the productive capacity of an economy need not necessarily be the full employment level of national product. We shall consider these issues in the chapters that follow.

REVIEW QUESTIONS

1 Explain why the relationship between price level and the size of national product (Q_{NP}) is downward-sloping when Q_{NP} is measured as the sum of expenditures, and upward-sloping when Q_{NP} is measured as the sum of incomes.

2 List the factors that influence shifts in AD and AS and give at least three examples of how governments have used these factors to manage the economy.

3 Australia underwent a recession during the following years: 1974–75, 1980–81, and 1988–89. Explain the causes that prompted the recession during these periods using the AD–AS frameworks described in this chapter.

11

Environmental Capital

Investment and Depreciation

The main focus of this chapter is to set the stage for the internalisation of environmental capital into macroeconomic analyses, so we shall revisit the concept of environmental capital that was introduced in the first two chapters of this text. In the previous chapter, three measures of economic performance—GDP, GNP, and NNP—were introduced. Although NNP is the better measure of the three in terms of depicting the performance of an economy, it does not account for the services of the natural environment. Therefore, the basis for internalising the environment into macroeconomic analyses rests primarily on modifying the definition of NNP to capture the services provided by the natural environment for the functioning economic systems.

However, one should note that the 'performance of an economy' is considered here (and in most texts) within the narrow scope of measuring the size of national product. No consideration is given to the various factors that influence the quality of life in a nation. Although modifying NNP for environmental services could go a long way towards addressing some quality-of-life issues, many others will remain unresolved.

Environmental capital

To enable the modification of NNP, it is useful to conceptualise the natural environment in the aggregate (that is, in its totality including all endowments—biological and physical) as a capital asset. Then the natural environment can be treated in national income accounting in the same way as traditional capital assets, such as buildings and machinery. That is, in the same way as national income statisticians compute capital stock estimates for manufactured capital, one needs to envisage the existence of a natural capital stock. We shall denote this as KN. Then, as illustrated below, any additions to this stock will represent investment, while diminution of this stock will represent depreciation.

Although the terms 'environmental capital' and 'natural capital' have now come into widespread use (Mäler 1991; Hartwick 1990; and Dasgupta, Kriström, & Mäler 1994), the concept of environmental capital is not necessarily recent. Recall the references in chapter 1 to several writers such as Marshall (1891), Gray (1914), Schikele (1935), Hotelling (1949), Ciriacy-Wantrup (1938), and Bunce (1942), who dealt with nature as capital in their analyses. Alfred Marshall's (1891) text (which is perhaps the first concise book on modern neoclassical economics) describes only *nature* and *man* as the principal agents of production. About the same time as Marshall was Professor of Economics at the University of Cambridge in England, Irving Fisher, who was a professor in economics at Yale University in the United States, published a seminal paper (1904) that has served as the cornerstone for studies on capital markets. In his review of the definitions of capital, Fisher drew on environmental assets such as lakes and rivers as analogies to explain the concepts of stocks and flows. In his attempt to lay out a conceptual framework for capital, Fisher took it for granted that nature was capital.

Capital goods are usually distinguished from others in terms of three related features:

- durability (they last over a long period of time)
- provision of services over time
- depreciation (reduction in the ability to provide services over time).

It is not difficult to envisage the natural environment in these terms. Like any other form of capital, nature is durable, and it generates a flow of services over time. It also degrades with use. One may argue that the durability of KN could be prolonged over a very long time horizon if human interference in terms of rearranging nature were kept to a minimum. As indicated in chapter 2, the durability of KN is manifested in the three types of interrelated services that it provides, namely: supplying raw materials ranging from the air we breathe to the minerals we extract, being a receptacle for the wide array of waste emissions and discharges, and providing aesthetic amenities such as landscape and scenery.

Distinguishing environmental investments from depreciation

In any economy, two types of activities, as illustrated in table 11.1, can be undertaken in terms of the flow of services from KN. The first category results in an increase in the total volume of services from nature, while the second attempts to maintain the existing flow of services provided by nature. Hence it is pertinent to regard the former as environmental investment and the latter as environmental depreciation.

Table 11.1 Environmental investment versus environmental depreciation

Environmental investment	*Environmental depreciation*
Activities that are designed to restore the flow of services from endowments that have **ceased to provide services**	Activities that are designed to maintain the flow of services from endowments that **currently provide services**
Analogy—replacement in the context of an obsolete investment	Analogy—expenditures to offset wear and tear
Examples:	Examples:
• Creation of wetlands for denitrification	• Pollution control activities by the public sector—municipal waste treatment
• Reforestation of areas that have been cleared for years of open-cut mining	• Maintaining the flow of surface and groundwater systems
• Restoring rivers rendered unusable due to algal blooms	• Maintaining the stocks of biological resources—forests, fish stocks
• Detoxifying unusable soils for urban or rural development	

Environmental investments

Environmental investments may be explained by recourse to some examples. Consider the Baltic Sea. Nearly half of it below a certain depth is believed to be a dead resource. A range of anthropogenic factors, including pollution emissions and over-exploitation, have contributed to a lack of dissolved oxygen. In a word, this part of the Baltic Sea is at present incapable of generating any service. If expenditures are incurred to restore the flow of services, then these restoration expenditures represent a new investment. Similarly, when a badly polluted beach (say, one that no one dares go to) is cleaned up, then the beach becomes capable of generating a flow of services. The analogy is building a new factory when the old one has become obsolete and unproductive. In traditional national income accounting, new investments are included as gross investments. Likewise, investment-type expenditures for restoring lost (obsolete) endowments can be included in the accounting framework as a new gross investment.

The controversy in the literature concerns whether the restoration expenditures should be included in NNP, and centres around this question: Are we really adding to the stock of environmental capital? The restored endowments were already there in the first place. The counter argument is as follows. If a river is dead due to the presence of a variety of pathogens, then that river cannot be counted in the stock of environmental capital, since it will not contribute to national output.

Removing the pathogens and restoring the river is tantamount to adding to the stock of capital that contributes to national output. One cannot of course ignore the fact that what is restored will not be authentic in terms of the item of KN that became obsolete. That is, the river we get back will most probably be different from its pristine state. Nevertheless, one needs to acknowledge that the restoration of the river results in an increase in the total volume of services from nature regardless of the issue of authenticity. Hence, the stock of KN at any given time will be made up of:

- naturally occurring endowments
- lost endowments that have been restored by human effort.

When the latter are additions to the prevailing stock of KN, they can be regarded as investments. Further, if all lost (obsolete) endowments of an economy have been restored, then there is no scope for environmental investments, and accounting procedures will be confined to the depreciation of the capital stock.

Allowance for depreciation of environmental capital

The second category of activities in table 11.1 deal with offsetting the wear and tear of KN. Thus, the expenditures due to them should be deducted from NNP. Sometimes, the distinction between environmental investments and the allowances for depreciation can be confusing. The following examples will help clarify the issues:

1   Consider again a polluted river. But in this case the river currently provides services while it is also becoming contaminated. The ongoing decontamination of the river, to maintain its role in contributing to national income, is similar to capital consumption or replacement investment. This is different from restoring a river that was not providing any services and was regarded as obsolete.

2   In the context of mineral resources, any new discovery will add to the stock of mineral wealth, and accordingly, resource discovery is investment. Following the general theory of the mine (Hartwick & Olewiler 1986), the depreciation of a mineral resource can be explained by depletion and the reduction in the value of the mine. This reduction has been usually explained in terms of the concept of user costs, namely the loss of consumption benefits by future generations.

Sustainable income

If, during a given time period t, net national product, environmental investments, and environmental depreciation are respectively denoted by NNP(t), $I_g(t)$, and $C_{EM}(t)$, then sustainable income would be defined as: $\{NNP(t) + I_g(t) - C_{EM}(t)\}$. If we can suppose that current estimates of NNP(t) include the expenditures on environmental investments, then the definition of sustainable income is simply $\{NNP(t) - C_{EM}(t)\}$. In fact most expenditures pertaining to environmental investments are included in the estimates of NNP, while this is not the case with environmental depreciation costs, as explained in the next section below.

Following Hotelling's (1931) and Keynes' (1936) expositions of permanent income from a capital good, the adjusted value of national income can be sustainable if at least two conditions are satisfied. These are:
- there is no diminution in the stock of environmental capital
- the value of environmental depreciation is less than the rent generated by the stock of environmental capital.

A steady state can be defined as one where:
- all lost endowments have been restored by way of investments
- positive returns net of depreciation are being maintained.

NNP(t) can be regarded as the economic return from a nation's environmental capital. Sustainable income is, hence, the difference between this economic return NNP(t) and the allowance made for the depreciation of environmental capital. We have defined the depreciation allowance as C_{EM}. Therefore, when the amount a nation spends exceeds $\{NNP(t) - C_{EM}(t)\}$, then that expenditure is not sustainable.

Recall from chapter 10 the definition of NNP, namely $(C + I + G + X - M - R - K_C)$. An examination of the national income accounts of most countries is likely to reveal that some costs of environmental depreciation appear as positive items in NNP; that is, they are mostly included in either C or I or G of NNP. For example costs of waste management are normally found in G, and expenditures by firms and households in terms of water filters and air filters to offset the deterioration of amenities are included in C. At the same time, other items such as the loss of topsoil and biodiversity are ignored. Hence in the estimation of $C_{EM}(t)$ one needs to separate out items that are currently included in NNP, as well as determine values for items that have been ignored. The following classification could prove useful in identifying the components of $C_{EM}(t)$:

1   *Costs involving production*—These are expenditures incurred by producers to maintain the services of the environment. They include the costs of complying with pollution control regulations and, in general, the costs of treating and disposing of the wastes that are generated from production. Most of these expenses are likely to be included in NNP (as positive items) under G.

2   *Costs involving current consumption*—These refer to expenditures that attempt to enhance the safety of consuming environmental services; for example water filters on taps and air filters on ventilation shafts. These could also include medical expenses due to illnesses induced by polluted environments. These expenses too are likely to be included in NNP as positive items under C or I.

3   *Costs involving future consumption*—These costs arise from the depletion of the stocks of renewable and non-renewable resources and the imposition of costs on future generations by the non-availability of these resources. The loss of biodiversity belongs to this category, and most of these costs are largely ignored.

A conceptual framework for $C_{EM}(t)$

Consider now the relationship between $C_{EM}(t)$ and national income $Y(t)$. It is reasonable to assume that increases in $Y(t)$ would prompt increases in $C_{EM}(t)$, and that increases in $Y(t)$ are feasible only up to a threshold. Any attempt to increase $Y(t)$ beyond this threshold could push $C_{EM}(t)$ towards infinity. This is due to the loss of assimilative capacity and irreversible damages that would be inflicted on the environment. This is consistent with the specific types of production functions and isoquants that were proposed in chapter 4 in the context of utilising KN. Such a conceptualisation for $C_{EM}(t)$ places KN in the category of non-renewable resources; for example see McInerney (1981). Besides, as was argued in chapter 4, the treatment of KN as a non-renewable item is valid, given that it is a complex system of resources and that the ability to recoup the assimilative capacity from such a system is inevitably finite.

Initially assume that the relationship between $C_{EM}(t)$ and $Y(t)$, namely $C_{EM}(t) = g\{Y(t)\}$, can be as described in figure 11.1, as follows:

$$C_{EM}(t) \quad = \quad C_{ER}(t) + \omega Y(t) \text{ for } \{0<Y(t)<Y_h(t)\}, (\omega>0) \tag{11.1}$$

$$C_{EM}(t) \quad \rightarrow \quad \infty \quad \text{for } \{Y(t)>Y_h(t)\} \tag{11.2}$$

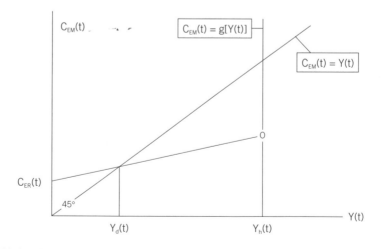

Figure 11.1   Environmental depreciation function—linear and discontinuous

That is, during any accounting period t, $Y_h(t)$ is the maximum limit to which output can be produced. This is a limit in terms of the assimilative capacity of the environment. Any attempt to increase income beyond $Y_h(t)$ results in irreversible environmental damage, and hence $C_{EM}(t)$ tends to infinity. Further, within the limit of $Y_h(t)$ following (11.1) above, the size of $C_{EM}(t)$ is governed by the extent of environmental maintenance that has to be done regardless the size of income, namely $C_{ER}(t)$; and ω, the rate at which $C_{EM}(t)$ increases for unit increases in Y. ω can be also regarded as the marginal rate of environmental degradation. That is:

$$\omega = \Delta C_{EM}(t)/\Delta Y(t) \tag{11.3}$$

Suppose that figure 11.1 represents the state of the environment for a specific accounting period. The feasible set of output targets for this period are hence defined by the domain $\{Y_d(t) < Y(t) < Y_h(t)\}$.

This is because over this domain the depreciation allowance for nature is below the 45-degree line; that is, $C_{EM}(t)$ is less than Y(t). The upper limit of this domain, $Y_h(t)$, denotes the target income that maximises Y(t) in excess of $C_{EM}(t)$. However, from (11.2), we can see that $Y_h(t)$ is also the income level that brings the economy to the brink of an environmental disaster. This difficulty is caused due to the linear and discontinuous assumptions that have been made for $C_{EM}(t)$ = g$\{Y(t)\}$ in (11.1) and (11.2). Therefore, a non-linear relationship such as an exponential function is perhaps more appropriate. Yet the linear function can prove useful, especially for joint consideration in linear macroeconomic models, and when we are able to assume that $Y_h(t)$ is sufficiently large to be ignored.

Consider now an exponential cost function of the following form:

$$C_{EM}(t) = C_{ER}(t)e^{\eta Y(t)} \tag{11.4}$$

In (11.4), η represents a compounding rate for the environmental costs. $C_{ER}(t)$ represents the specific value of $C_{EM}(t)$ when $\{Y(t) \to 0\}$. That is, as indicated before, it is the amount of environmental maintenance that has to be performed regardless of the size of national income. From (11.4) it is possible to define η as:

$$\eta = \{\ln C_{EM}(t) - \ln C_{ER}(t)\}/Y(t). \tag{11.5}$$

From figure 11.2, which represents this exponential function, it can be seen that the feasible set of output targets are given by the domain $\{Y_d(t) < Y(t) < Y_u(t)\}$. Further, note that in figure 11.2 output is feasible only when the depreciation function can intersect the 45-degree line, and this can in turn occur only when the gradient of the function is <1. Alternatively, when the gradient is >1, the environmental depreciation function is above the 45-degree line, indicating that the state of the environment is being heavily degraded. Recall that in the previous chapter we considered the concept of productive capacity. It is now possible to illustrate how KN can become a determinant of productive capacity. As is evident from figure 11.2, an economy loses its productive capacity if $C_{EM}(t)$ lies above the 45-degree line; that is, the gradient of $C_{EM}(t)$ exceeds 1.

The gradient of $C_{EM}(t)$ is also the marginal rate of environmental degradation and can be defined as:

$$dC_{EM}(t)/dY(t) \quad = \quad C_{ER}(t)\eta e^{\eta Y(t)} \tag{11.6}$$

The maximising value of national income, namely $Y_h(t)$ in figure 11.2, can be derived by equating (11.6) to 1. That is:

$$Y_h(t) \quad = \quad -\{lnC_{ER}(t) + ln\eta\}/\eta \tag{11.7}$$

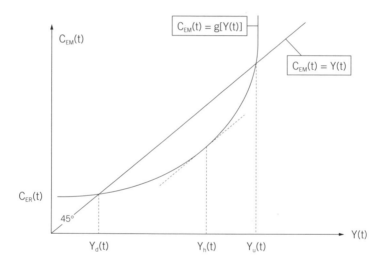

Figure 11.2   Environmental depreciation function—exponential

Some empirical evidence

The relationship $C_{EM}(t) = g\{Y(t)\}$ was tested as a time series comparison of national income in the United States with environmental depreciation data from Daly and Cobb (1989). The following result was obtained for the exponential function, when $C_{EM}(t)$ was equated to the sum of costs associated with water, air, and noise pollution and the loss of wetlands and farmlands:

$$ln\,C_{EM} \quad = \quad 4.94 + 0.00045\ NNP \tag{11.8}$$
$$(183.8) \quad (35.13) \quad r^2 = 0.972$$

A brief description of the Daly–Cobb data is as follows. Costs of water pollution were based on lost recreation benefits due to point source discharges and the costs of dredging operations against siltation. Air pollution costs included damages to agricultural vegetation, materials damage, and the effects of acid rain. These were estimated using hedonic prices and wage differentials. The costs of noise pollution were taken from estimates provided by the World Health Organisation. The loss of wetlands was valued on the basis of land values and environmental quality scores for attributes such as habitats and aesthetics. The loss of farmland due to soil degradation was valued in terms of lost agricultural output, and the loss of non-renewable resources was estimated as the forgone value of mineral output. Long-term environmental damage was assumed to be caused by non-renewable energy consumption, and was valued in terms of an energy consumption tax.

A similar test with Australian data (Thampapillai & Uhlin 1994), where $C_{EM}(t)$ was assumed to be made up of the expenditures of firms dealing with waste management, revealed the following relationship, which also qualifies the exponential form:

$$\ln C_{EM} = 4.0 + 0.005 \text{ GDP} \qquad (11.9)$$
$$\phantom{\ln C_{EM} = } (30) \quad (7.7) \quad r^2 = 0.72$$

In both cases, t-values (in parentheses) and r^2 are satisfactory.

Concluding remarks

To summarise, we have considered in this chapter the basis for modifying NNP such that we can derive an expression for sustainable income. This basis involved conceptualising nature as capital and drawing the distinction between environmental investments and the depreciation allowance for nature. A major challenge is the measurement of the environmental depreciation allowance. Empirical evidence seems to suggest that the relationship between this allowance {$C_{EM}(t)$} and national income is an exponential function. Until systems of environmental accounting are properly set up, the estimation of $C_{EM}(t)$ has to follow by the use of proxies. In the next chapter we shall consider the internalisation of the depreciation allowance of nature into selected macroeconomic models.

REVIEW QUESTIONS

1 Critically review the following statement: All expenditures incurred with respect to KN represent depreciation allowances and it is not possible to justify any environmental expenditure as an investment.

2 Explain how $C_{EM}(t)$ could become a determinant of productive capacity.
3 Suppose that the following data are available for a specific economy:

$C_{EM}(t) = 50$ units; $Y(t) = 2000$ units; $C_{ER}(t) = 5$ units

Estimate whether this economy is within its productive capacity with reference to KN.

12

The Environment and Income Determination

In this chapter, we shall illustrate the effects of internalising environmental capital into aggregate demand. For this purpose we shall use an introductory Keynesian framework of income determination. The main assumptions for this framework are that output is determined by aggregate demand and the price level is constant. The first assumption implies that the aggregate supply curve is horizontal, as shown in figure 10.2A in chapter 10. The simple Keynesian framework is chosen because it is, perhaps, the cornerstone of the various income and price determination models that have evolved over time.

The income determination framework is illustrated in three steps. In the first, we consider income determination without any reference whatsoever to environmental capital. In the second step, we shall introduce environmental capital and assume that the function describing environmental depreciation is linear. Understandably, this type of description poses some difficulties, due especially to environmental irreversibilities. Therefore, in the third step, environmental depreciation will be explained in terms of a non-linear exponential function along the lines suggested in chapter 11.

Equilibrium income without environmental depreciation

In a given time period, t, the equilibrium between the aggregate demand {NNP(t)} and income Y(t), will be:

$$Y(t) \equiv NNP(t) \tag{12.1}$$

For reasons of simplicity assume that all components of NNP except consumption are constant. We denote this constant by $\Phi(t)$—that is: $\{\Phi(t) = I+G+X-M-R-K_c\}$.

In most macroeconomic texts, consumption in a given year $C(t)$ is usually defined as a linear function in the following way:

$$C(t) = \alpha(t) + \beta(t)Y(t) \qquad (12.2)$$

where $\alpha(t)$ is *autonomous consumption*, namely the consumption we need regardless of our income; and $\beta(t)$ is the amount by which our consumption will increase when income increases by one unit; this is more commonly known as the *marginal propensity to consume*. We shall assume, again for convenience and simplicity, that $\alpha(t)$ is contained in G of $\Phi(t)$. So $NNP(t)$ can be written as:

$$NNP(t) = \Phi(t) + \beta(t)Y(t) \qquad (12.3)$$

The value of income that satisfies the identity $[Y(t) \equiv NNP(t)]$ is equilibrium income $Y(t)^*$, and this can be found by substituting equation (12.3) into (12.1) and solving for Y:

$$Y(t) \equiv \Phi(t) + \beta(t)Y(t)$$

$$Y(t)\{1 - \beta(t)\} = \Phi(t)$$

$$Y(t)^* = \Phi(t)/\{1 - \beta(t)\} \qquad (12.4)$$

In figure 12.1, this equilibrium income is determined by the point of intersection of the schedule describing $\{NNP(t) = \Phi(t) + \beta(t)Y(t)\}$ with the 45-degree line. This 45-degree line describes all points where $\{Y(t) = NNP(t)\}$.

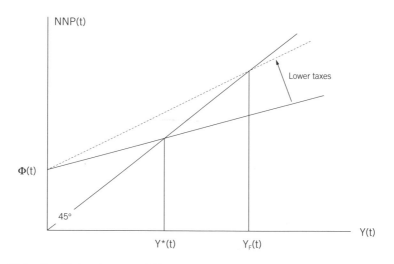

Figure 12.1 Equilibrium income within an elementary linear Keynesian model

Suppose now that national income has to be raised from $Y(t)^*$ to $Y_F(t)$ because policy-makers believe that they can achieve full employment at an income level of $Y_F(t)$. Since we have assumed that output (income) will be determined by aggregate demand, then it is logical to stimulate aggregate demand, say by lowering taxes and/or increasing government spending. This stimulus will result in an upward shift of the aggregate demand schedule, as shown in figure 12.1. Thus we shall observe how policy-makers can achieve specified income targets through manipulating shifts in aggregate demand. This is referred to as 'pump-priming'. As we shall see below, pump-priming takes on an added dimension when we internalise the depreciation of environmental capital into aggregate demand.

It is also important to note that the quantity $[1/\{1 - \beta(t)\}]$ in equation (12.4) is referred to as the 'multiplier' in standard macroeconomic theory. That is, if aggregate demand is raised by one unit, then income will increase by an amount equal to the multiplier. As we will show below, recognising the depreciation of nature makes the size of the multiplier smaller.

Equilibrium income with environmental depreciation—linear

We start our analysis with the equilibrium between the aggregate demand and sustainable income. For a given time period, t, this will be defined, following the definition given in chapter 11, as:

$$Y(t) \equiv NNP(t) - C_{EM}(t) \tag{12.5}$$

We have already defined $NNP(t)$ as $\{\Phi(t) + \beta(t)Y(t)\}$ in equation (12.3) above. In defining the depreciation allowance for environmental capital, we make two further simplifying assumptions in relation to the definitions offered in chapter 11. First we assume that the upper limit for income that renders the depreciation allowance to be infinite is large enough (relative to current income) to be ignored. Second, we also assume that the amount of environmental maintenance that needs to be done regardless of income (namely C_{ER} in equation (11.1) in chapter 11) is zero. These assumptions enable the depreciation allowance for environmental capital, $C_{EM}(t)$, to be defined as a linear proportion, $\gamma(t)$, of NNP:

$$C_{EM}(t) = \gamma(t)\{NNP(t)\} = \gamma(t)\Phi(t) + \gamma(t)\beta(t)Y(t) \tag{12.6}$$

The value of sustainable equilibrium income will be defined as:

$$Y(t)^{**} = [\Phi(t)\{1 - \gamma(t)\}] / [1 - \{\beta(t)\{1 - \gamma(t)\}\}] \tag{12.7}$$

As one would expect, the magnitude of $Y(t)^{**}$ is directly proportional to $\Phi(t)$ and $\beta(t)$, and is inversely proportional to $\gamma(t)$. The determination of $Y(t)^{**}$ is illustrated in figure 12.2.

In an economy portraying the characteristics of figure 12.2, if $C_{EM}(t)$ is ignored, then income determination will be based on NNP alone. As illustrated in figure 12.2, the expenditure schedule {NNP(t)} results in income being determined as $Y(t)^*$. Hence, the difference {$Y(t)^* - Y(t)^{**}$} may be regarded as the current consumption of income at the expense of maintaining a permanent flow of income. Setting aside the allowance for the depreciation of environmental capital—namely incurring $C_{EM}(t)$—maintains environmental capital and thus results in a smaller but steady flow of income.

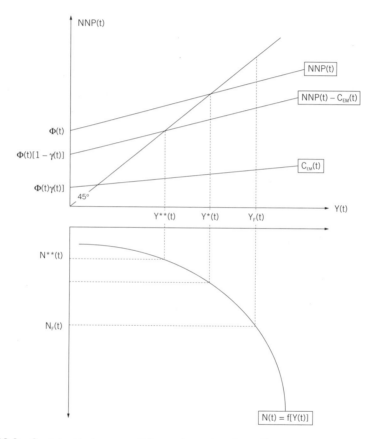

Figure 12.2   Sustainable income within an elementary linear Keynesian model

The recognition of $C_{EM}(t)$ also reduces the size of the multiplier; that is:

$$\text{from } [1/\{1 - \beta(t)\}] \text{ to } [1/\{1 - \beta(t)(1 - \gamma(t))\}]$$

Should $Y_F(t)$ be the capacity income for achieving full employment, then the standard Keynesian approach would be to pump-prime the economy by raising the gradient of aggregate demand or shifting it upwards, as illustrated on page 149, for example by increased government spending with some faith in the theory of multipliers. However, note that this upward shift may also be achieved by lowering $\gamma(t)$; that is, by prompting a downward shift in the schedule describing $C_{EM}(t)$.

There are basically two broad approaches for achieving a downward shift in the $C_{EM}(t)$ schedule. The first is a technology-based solution: to find, if one might coin a phrase, *environment-saving* technologies. These could include a very wide range of options involving cost-effective methods of utilising the environment for economic activities such as: advances in molecular biology to remove algal blooms, improved methods of waste treatment and recycling, and the use of solar energy for domestic and commercial uses. The second is a lifestyle option in terms of managing with less income rather than with more. In figure 12.2 we can also observe the relationship between income and employment. That is, economic agents, whether they be producers or consumers, adopt conscious decisions to moderate their behaviour and thereby inflict less harm on environmental capital. To achieve full employment, $N_F(t)$, we need an income of $Y_F(t)$. However, with the sustainable equilibrium income $Y(t)^{**}$, we can employ only $N(t)^{**}$ persons, and this implies that $\{N_F(t) - N^{**}\}$ persons will be unemployed.

Suppose that $Y_F(t)$ and $Y(t)^{**}$ are respectively $600 billion and $300 billion and that $N_F(t)$ and $N(t)^{**}$ are respectively 3 million and 2 million. With a national income of $Y_F(t) = $600 billion, in the context of full employment, average income will be $20,000 per head. With a much lower sustainable income of $300 billion and an unemployment level of 1 million persons, the average income per head (among those employed) will be $15,000. Should this income be shared by the entire labour force so as to have full employment, then the average income per head reduces to $10,000. That is, to enable full employment as well as the sustainability of national income, workers would need to make a sacrifice of $10,000 per head, if their expectations were centred around the higher national income of $600 billion. If they reconciled with a sustainable income of $300 billion, then the sacrifice to be made for full employment would be $5,000 per head. Such wage sacrifices represent a lower wage policy, which entails a range of lifestyle changes. These include using public transport instead of private transport, fewer energy-intensive facilities, and in general a curbing of extravagant consumption. To illustrate: in the northern hemisphere people generally lounge around in sleeveless shirts in midwinter with the thermostat set at around 25°C. Other extravagances include the proliferation of devices that substitute for human effort, such as electric toothbrushes and carving knives.

A difficulty with this framework is that it is possible to achieve $Y_F(t)$, even if $\gamma(t)$ were not lowered, by simply increasing $\Phi(t)$; that is, by prompting an upward shift in the expenditure schedule. The implication of this is that higher levels of national income can be attained so long as the correspondingly higher levels of $C_{EM}(t)$ are also incurred. This can be misleading, since it overlooks the inherent irreversibility of environmental changes. For this reason, a non-linear framework is considered below.

Non-linear framework

When irreversible environmental changes occur, $C_{EM}(t)$ will tend to increase at a rate greater than a fixed rate as implied by a linear function. Hence we shall use the exponential function that was proposed in chapter 11:

$$C_{EM}(t) \;=\; C_{ER}(t)e^{\eta(t)[NNP(t)]} \;=\; C_{ER}(t)e^{\eta(t)[\Phi(t)+\beta(t)Y(t)]} \tag{12.8}$$

Recall from (11.5) that $\eta(t)$ is the proportion $[\{\ln C_{EM}(t) - \ln C_{ER}(t)\} /[NNP(t)]]$. With the exception of (12.3) being replaced by (12.8), all other relationships described above are assumed to remain intact. Thus the relevant expressions to define the economy are: (12.1), (12.2), and (12.8), and sustainable output can be defined as:

$$Y(t) \;=\; \Phi(t) + \beta(t)Y(t) - \{C_{ER}(t)e^{\eta(t)[\Phi(t)+\beta(t)Y(t)]}\} \tag{12.9}$$

For a given value of η, a positive equilibrium is feasible, so far as values of NNP exist such that $\{C_{ER}(t)\eta e^{\eta(t)[NNP(t)]} < 1\}$. The determination of equilibrium income in figure 12.3 could involve a computational approach; for example iteratively changing the value of $Y(t)$ until the LHS of (12.9) equals its RHS.

The main difference between the linear and non-linear frameworks concerns the extent to which pump-priming can continue by prompting upward shifts of

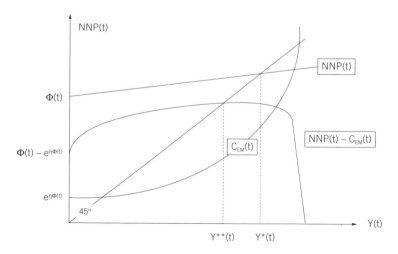

Figure 12.3 Non-linear system and sustainable income within a Keynesian model

the components of final demand. With the linear framework, pump-priming can continue indefinitely, so long as the gradient of (NNP–C_{EM}) does not exceed 1. With the non-linear framework, prompting the upward shift, for example by raising $\Phi(t)$, could at some point shift the C_{EM} schedule above the 45-degree line, and thereby render a positive equilibrium unfeasible.

Income determination and policy analyses

The linear and non-linear formulations of income determination frameworks can be empirically applied for policy analyses when the valuation of C_{EM} is possible. The review of the literature on environmental accounting reveals that several proxies have been used for C_{EM}. These include:

1 total energy consumption as a proxy for costs of pollution from using energy resources (Thampapillai & Uhlin 1996)
2 cost of fertiliser application for soil erosion (Repetto & Magrath 1989)
3 stumpage value of forests for deforestation (Repetto & Magrath 1989; Cruz & Repetto 1992)
4 market value of fish for over-fishing (Cruz & Repetto 1992)
5 loss in agricultural output for soil degradation (Young 1992)
6 a range of proxies for quality-of-life considerations, as indicated in chapter 11 (Daly & Cobb 1989).

To illustrate the analyses of some specific policy questions, we choose the USA as a case study and use the cost of energy consumption as a proxy for C_{EM}. The use of this proxy can be defended on the grounds that energy is a basic input in all production, and that its utilisation is the dominant source of pollution. The C_{EM} data derived by this proxy method for the US economy are illustrated in table 12.1. Because the macroeconomic aggregates used for our illustration were estimated in 1982 dollars, the total consumption of energy in coal equivalents was multiplied by the 1982 price of coal. Table 12.1 also contains the estimates of C_{EM} from Daly and Cobb (1989, pp. 418–19), namely the sum of their costs for air, water, and noise pollution, and the loss of wetlands and farmlands. As can be seen, the energy expenditure proxy values are reasonably close to those of Daly and Cobb. In fact, a t-test qualifies a lack of significant difference between the two sets of values at the 95 per cent confidence level. The policy analysis presented here is taken from Thampapillai and Uhlin (1996, 1997).

For the purposes of illustrating policy analyses, we consider three questions:

• Is the US economy showing some indication of a movement towards attaining sustainability?
• To what extent must wages be reduced in order to achieve sustainability?
• To what extent must the efficiency of the natural environment be improved for achieving sustainability and can we make investments in the environment?

Table 12.1   Energy resource expenditure in the US economy compared with residual damage expenditures from Daly and Cobb

Year	Energy input coal equivalent	Energy input expenditure (1982 prices)	Environmental damage cost from Daly & Cobb
	Tonnes 10^9	$ billion	$ billion
1976	2.5419	158.3922	164.00
1977	2.5239	157.2712	166.01
1978	2.7615	172.0749	166.37
1979	2.6294	163.8454	168.02
1980	2.6472	164.9575	167.65
1981	2.4752	154.2356	168.39
1982	2.3250	144.8749	168.01
1983	2.3992	149.4994	168.56
1984	2.4721	154.0469	171.66
1985	2.4880	155.0349	173.47
1986	2.4827	154.7033	175.30
1987	2.5302	157.6616	177.97
1988	2.6935	167.8424	180.73
1989	2.8216	175.8249	183.56
1990	2.7935	174.0714	186.49
1991	2.6989	168.1778	189.49
1992	2.7956	174.2027	192.59

The data in Daly and Cobb were for the period 1950–86. The values given in this table for 1987–92 were drawn by extrapolation.

Source: World Development Report (several issues); Daly and Cobb (1989)

Before we proceed any further, we must note that our definition of sustainability is rather narrow (compared to what ecologists might profess) and is confined to (NNP–C_{EM}).

Indications of achieving sustainability

To analyse the first question, we can estimate trends for the coefficients that determine Y* and Y** and project the time paths of these equilibrium income values. Recall that Y* represents the equilibrium that is based on standard macroeconomic data—that is (Y ≡ NNP), while Y** represents the sustainable equilibrium (Y ∫ NNP ≡ C_{EM}). If the projected paths of Y* and Y** are showing signs of convergence, then it is possible to conclude that sustainability considerations are being included in standard macroeconomic practices. The coefficients for which trends need to be estimated are those specified in equations (12.4), (12.7), and (12.9), namely $\Phi(t)$, $\beta(t)$, $\gamma(t)$, and $\eta(t)$. Thampapillai and Uhlin (1997) estimated these using single period lag models and obtained the following results:

$$\Phi(t)) \quad - \quad 1.011\Phi(t-1) \quad R^2 \ = \ 0.99$$
$$(73.89) \qquad\qquad D \ = \ 2.036 \qquad\qquad\qquad (12.10)$$

$$\beta(t)) \quad = \quad 1.0023\beta(t-1) \quad R^2 \ = \ 0.99$$
$$(309.74) \qquad\qquad D \ = \ 2.513 \qquad\qquad\qquad (12.11)$$

$$\gamma(t)) \quad = \quad 0.981\gamma(t-1) \quad R^2 \ = \ 0.99$$
$$(105.44) \qquad\qquad D \ = \ 3.06 \qquad\qquad\qquad (12.12)$$

$$\eta(t) \quad = \quad 0.977\eta(t-1) \quad R^2 = 0.99$$
$$(180.26) \qquad\qquad D = 2.12 \qquad\qquad\qquad (12.13)$$

(Numbers in parentheses represent t-values and D stands for the Durban-Watson statistic.)

Recall that γ represents the coefficient for environmental depreciation in the linear model, while this coefficient for the non-linear model is represented by η. In order to illustrate the significance of the changes in $\gamma(t)$ and $\eta(t)$, the analysis was performed as follows:

1 Sustainable income paths were projected by the application of equations (12.4) and (12.7) for each year of the period 1980–91 by holding the values of $\gamma(t)$ and $\eta(t)$ constant at the 1980 level, but letting $\Phi(t)$ and $\beta(t)$ change according to the lag relationships shown above.

2 The analysis was then repeated, but this time by also allowing $\gamma(t)$ and $\eta(t)$ to change in line with the lag relationships. That is, in the derivation of these new income paths, $\gamma(t)$ and $\eta(t)$ would become progressively smaller each year— although at a very low rate.

The sustainable income path from each application above was then compared with the following three values: *actual realised NNP, capacity NNP required for full employment*, and *NNP based on the equilibrium {Y=NNP(t)}* (table 12.2 and figures 12.4 and 12.5). In each comparison, convergence tests were also performed. The following observations emerge.

1   If $\gamma(t)$ or $\eta(t)$ remains constant, there is no possibility of convergence between the actual NNP path and the sustainable paths. That is, the difference (actual NNP–Y^{**}) increases over time.

2   When $\gamma(t)$ and $\eta(t)$ are permitted to reduce in line with trends derived from the time series data, there is a difference in the sustainable income paths. First, the paths derived from the non-linear framework now reveal a faster growth in income than the paths derived from the linear framework. Second, there is a possibility of convergence between the actual NNP path and the sustainable paths. This possibility appears stronger for the paths derived from the non-linear framework.

Table 12.2   Comparison of sustainable, actual, and capacity incomes (1980–91) (all values in 1982 US dollars billion)

Year	Y*1L	Y*1NL	Y*2L	Y*2NL	Y*	Actual NNP	Net capacity Y
1980	2668.10	2668.02	2668.10	2668.02	2832.44	2832.57	3061.59
1981	2723.38	2732.00	2712.44	2690.24	2886.26	2878.59	3131.77
1982	2779.86	2795.10	2757.65	2709.64	2941.36	2782.81	3103.97
1983	2837.57	2859.50	2803.74	2728.32	2997.79	2897.40	3221.62
1984	2896.54	2925.22	2850.72	2746.26	3055.59	3115.91	3371.30
1985	2956.81	2992.30	2898.62	2763.48	3114.79	3278.01	3451.54
1986	3018.41	3060.73	2947.46	2779.94	3175.44	3313.71	3542.97
1987	3081.38	3130.60	2997.27	2795.63	3237.59	3390.81	3614.72
1988	3145.75	3194.58	3048.07	2810.56	3301.28	3527.85	3733.47
1989	3211.57	3272.22	3099.87	2824.72	3366.58	3625.21	3826.89
1990	3278.88	3346.60	3152.71	2838.13	3433.51	3644.93	
1991	3347.71	3424.96	3206.61	2850.76	3502.16	3578.45	

Y*1L—equilibrium income from linear model with changing γ; Y*2L—equilibrium income from linear model with constant γ; Y*1NL—equilibrium income from non-linear model with changing η; Y*2NL—equilibrium income from non-linear model with constant η. For more information about the data and methods used here, see Thampapillai and Uhlin (1997).

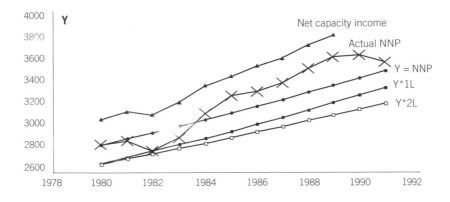

Figure 12.4   Comparison of income paths (1980–91)—linear model

Y*1L—equilibrium income from linear model with changing γ; Y*2L—equilibrium income from linear model with constant γ; Y*1NL—equilibrium income from non-linear model with changing η; Y*2NL—equilibrium income from non-linear model with constant η. For more information about the data and methods used here, see Thampapillai and Uhlin (1997).

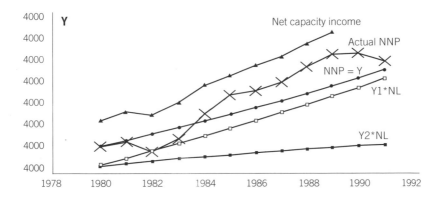

Figure 12.5   Comparison of income paths (1980–91)—non-linear model

Y*1L—equilibrium income from linear model with changing γ; Y*2L—equilibrium income from linear model with constant γ; Y*1NL—equilibrium income from non-linear model with changing η; Y*2NL—equilibrium income from non-linear model with constant η. For more information about the data and methods used here, see Thampapillai and Uhlin (1997).

3   Finally, there is virtually no possibility of convergence between the sustainable path and the income path for capacity output, regardless of the changes in γ(t) or η(t).

The observed differences between when $\gamma(t)$ or $\eta(t)$ are held constant and when they are allowed to change represent likely evidence of technological and institutional improvements in the management of environmental capital. These improvements are manifested in the rightward shift of the $C_{EM}(t)$ function over time. That is, efficiency improvements act to reduce the amount of environmental depreciation per unit of output.

However, some caution needs to be exercised with the interpretation of the above results for two reasons. First, the results rest essentially on environmental data, which were inevitably estimated by indirect methods of valuation. Second, the equilibrium incomes determined here are primarily short-run values due to the assumption of a fixed price level. Hence, the possible convergence between actual and sustainable paths should not be taken to imply that income growth can be indefinitely sustained.

Improving the efficiency of nature, investments, and wages

We now turn to the remaining two policy questions. Within the confines of the framework considered here, it is possible to examine the role of environmental technology, environmental investment, and real wages. For example it is possible to raise income by improvements in the 'efficiency of environmental capital use'. As indicated, these include developments such as cost-effective methods of waste treatment and better pollution control. Such improvements would be manifested in reductions of C_{EM} and $\gamma(t)$ or $\eta(t)$. Table 12.3 shows the extent to which $\eta(t)$ could have been reduced in order to make sustainable income equal to capacity income (that is, full employment income—Y_F) for the period 1987–89.

The relationship between improvement in environmental capital efficiency (ECE) and output is illustrated in figure 12.6 for each year from 1987 to 1989. This relationship has been derived from the non-linear framework by parametrically reducing $\eta(t)$, while keeping $\Phi(t)$ and $\beta(t)$ fixed at the levels observed for

Table 12.3 Required improvements in environmental capital efficiency for full employment (non-linear framework)

Year	1987	1988	1989
Observed η	0.001492	0.001452	0.001426
η for full employment	0.001157	0.001158	0.001146
Percentage reduction in η required for full employment	22.4256	20.2364	19.6613

the relevant years. For example, for the schedule describing the relationship for 1987, $\Phi(t)$ and $\beta(t)$ are fixed at the respective observed values of $889 billion and 0.773. ECE is measured in terms of the percentage reduction in $\eta(t)$. The upward shift in the income schedule between successive years is solely due to the increase in $\Phi(t)$. An examination of the US national income accounts reveals that between 1987 and 1989, increase in investment was by far the most significant contributor to the increase in $\Phi(t)$.

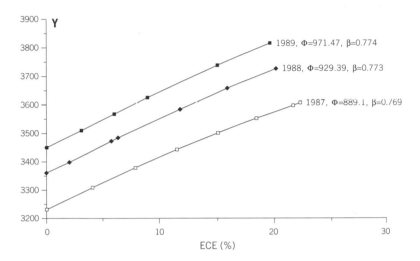

Figure 12.6 Environmental capital efficiency

Although the national accounts do not reveal sufficient disaggregation, it is possible to envisage a special class of investments termed 'environmental capital investments'. These can be important shifters of the income schedule. As discussed in chapter 11, they can be defined as the restoration of lost endowments—analogous to replacement investments in the context of sunk costs. To recapitulate an example, if at the start of a given period a river is declared dead owing to a variety of pathogens, then this river cannot be counted in the stock of environmental capital because it will not contribute towards national output. That is, its status will be similar to that of a sunk cost. Removing the pathogens and restoring the river is tantamount to adding to the stock of environmental capital. Hence, activities such as reforestation of mined-out areas, the creation of wetlands in water courses destroyed by nitrogen putrefaction, and the detoxification of contaminated land to enable new development are examples of environmental investments. Since these investments are reproducible capital, they can be regarded as special cases of the 'Hartwick Rule' in practice (Hartwick 1977, 1978; Solow 1986)—that is, making reinvestments in reproducible forms of capital to maintain a permanent flow of income.

As we illustrated on page 151, an alternative approach for achieving convergence between sustainable and capacity incomes is to lower real wages. For example consider 1989. The observed average annual real wage in the US economy then was $30,894 (in 1982 dollars) and the observed value of η(t) was 0.001426. Given this value of η(t) the value of equilibrium income is $3,449 billion. If this equilibrium income was to have provided full employment to the then labour force of 123.87 million, real wages needed to have fallen from $30,894 to $27,846. That is, those employed would have had to sacrifice $3,048 to draw the unemployed into the workforce. Should the value of η(t) have fallen (depicting an improvement in environmental capital efficiency), then the sacrifice to be made of real wages for increasing employment would also have fallen.

Therefore, we can envisage a trade-off between 'wage sacrifices for sustainability' (WS) and improvements in environmental capital efficiency (ECE). Such a trade-off schedule is illustrated in figure 12.7. Note that along a given trade-off schedule Φ(t) and the price level are fixed. Increases in Φ(t) (due, say, to environmental investments) and/or reductions in the price level will shift the trade off-curve inwards, implying the need for lower sacrifices of real wages. That is, in figure 12.7, the optimal situation is defined by the origin because it represents zero wage sacrifices and no further need for improvements in ECE.

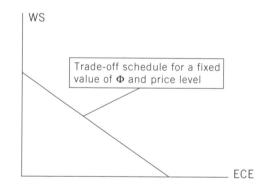

Figure 12.7   Trade-offs between ECE and WS

Some concluding remarks

Although we introduced a simple framework in this chapter, it was possible to illustrate some important policy applications. The assumption that aggregate supply is horizontal—that is, 'whatever is demanded is produced'—could seem to be restrictive. Nevertheless, this assumption can be applied within limits in economies that have a fair degree of unemployment. The important message from this chapter is that, during times of unemployment and a general economic recession, the natural environment can also be seen as a viable avenue for stimulating the economy. But policy-makers have traditionally opted for lowering taxes or interest rates or spending on items such as roads and other infrastructure, even when such items are in abundance. Alternatively, there are several options for reducing C_{EM} and as a result leading to increases in national income. For example reforestation provides bio-fuel, reduces energy imports, traps carbon from the air, and as result makes the workforce healthier. A healthier workforce produces more and thus raises income.

REVIEW QUESTIONS

1 Suppose that the following data are available for an economy:

$$\Phi = 241, \eta = 0.0083, \beta = 0.61, \text{ and } C_{ER} = 5$$

where Φ and C_{ER} are expressed in units of income.

- P3 155

Use a computational method and estimate the size of Y^{**} by applying the equality in equation (12.9) above.

Estimate the values of Y in equation (12.9) for various values of aggregate expenditure ranging from 10 units up to 500 units and plot the relationship between Y and expenditure. Graphically illustrate the value of Y^{**} in this plot.

Estimate the value of Y^* (equilibrium income when the depreciation of environmental capital is ignored).

2 If the economy has to achieve the value of Y^* (which you estimated above) for the purposes of full employment, estimate the amount by which η should be reduced. Discuss the practical ways by which a significant reduction in η can be achieved.

3 Discuss the feasibility of adopting a 'lower wages' policy in developed as well as developing countries.

13

The Environment in Other Macroeconomic Models

In this chapter we shall explore the avenues for internalising the natural environment into a further set of macroeconomic frameworks. As in the previous chapter, the conceptual premises that were developed in chapter 11 will be used as the basis for the internalisation. We will consider four frameworks here, namely the Harrod–Domar model of economic growth, the IS–LM model, a framework for deriving aggregate supply, and a framework of general equilibrium. We shall retain the assumptions of constant price level and a perfectly responsive (Keynesian) aggregate supply curve in the first two frameworks and remove them in the last two.

The Harrod–Domar (H–D) model of economic growth

We briefly reviewed this model in chapter 1. To recapitulate, consider first the derivation of the H–D growth model. Suppose that the capital stocks and income during a given period are denoted by K and Y, and these in the following period are (K + ΔK) and (Y + ΔY). Further, we denote the capital–output ratio as κ and assume that it remains constant. Then:

$$\kappa \ = \ K/Y \ = \ (K + \Delta K) / (Y + \Delta Y) \ = \ \Delta K/\Delta Y \qquad (13.1)$$

Hence it follows that the additions to capital stocks, which are also the investment (I) made during a period, can be defined as:

$$I \ = \ \Delta K \ = \ \kappa \Delta Y \qquad (13.2)$$

If savings (S) is defined by a proportion (ø) of national income, then assuming the savings–investment identity, that is (I ≡ S), it follows that:

$$I = S = øY = \Delta K \qquad (13.3)$$

Given that from (13.2) above ΔK is also equal to $(\kappa \Delta Y)$, the proportion of income saved can be expressed as:

$$øY = \kappa \Delta Y \qquad (13.4)$$

As we observed in chapter 1, the basic H–D model is the result of dividing both sides of (13.4) by κY. That is:

$$\Delta Y/Y = ø/\kappa \qquad (13.5)$$

The left-hand side of (13.5) is in fact the rate of economic growth: The inference from (13.5) is that the rate of economic growth is directly proportional to the savings ratio and is inversely proportional to the capital–output ratio. Hence economic growth will be accelerated if the capital–output ratio is low and the savings ratio is high. The policy implication from this simple analysis is the need to encourage savings and make investments in efficient forms of capital to achieve higher rates of economic growth.

Consider now the environmental cost function as explained in figure 11.1 and equations (11.1) and (11.2) in chapter 11. Although the non-linear environmental cost function is more appealing, the linear function is used here due to the discrete linear assumptions that are inherent in the H–D model. Suppose that figure 11.1 refers to a new accounting period. Given the environmental limits, the maximum output target in this period is Y_h; that is, $\Delta Y = Y_h$. The environmental costs that have to be incurred in excess of the initial maintenance costs to achieve this output target can be defined using (11.1) as:

$$(C_{EH} - C_{ER}) = \omega \Delta Y \qquad (13.6)$$

Dividing (13.6) by (13.4) and rearranging the resulting expression reveals that:

$$\Delta Y/Y = [(C_{EH} - C_{ER})/\Delta K][ø/\omega] \qquad (13.7)$$

This last statement is in fact a definition for the rate of economic growth in terms of the environmental cost function along with the additions to capital stocks and the savings ratio. Given that the rate of economic growth is set in terms of output targets, the definition in (13.7) can be used to determine the level of investment that will permit the achievement of the maximum feasible output target $(\Delta Y = Y_h)$. That is:

$$I = \Delta K = (C_{EH} - C_{ER}) (\kappa/\omega) \qquad (13.8)$$

From (13.8) it follows that the additions to capital stocks can be large when:
1   the gap between C_{EH} and C_{ER} is large, and/or

2   ω, the marginal rate of environmental degradation, is low.

Note that the first term on the right-hand side of (13.8) is Y_h, the maximum feasible income in terms of the environmental costs. Y_h is large when the gap between C_{EH} and C_{ER} is wide. Further, a very wide gap between C_{EH} and C_{ER} implies that the initial environmental maintenance costs can be low relative to the environmental cost that sets the limit to growth. Hence the potential to increase capital stocks diminishes as the costs approach the magnitude of the limiting costs. The results in (13.7) and (13.8) are not surprising; that is, to moderate investments when C_{ER} and/or ω are high. However, the analysis demonstrates that when an environmental cost function can be developed from the statement of environmental accounts, output targets can be nominated and the level of investments that comply with these output targets can be determined. Although the H–D type models deal with investments and savings, they do not directly account for the price of investment, namely interest rates. Hence, it is relevant to consider the IS–LM framework next.

The environment in IS–LM analysis

Following standard macroeconomic theory, the IS curve describes the relationship between the nominal interest rate (i) and national income (Y) in terms of savings that are invested. That is, as interest rates fall, people invest more, and these higher investments are manifested in higher levels of output. Hence the IS curve displays an inverse relationship between i and Y. The LM curve describes the relationship between i and Y in terms of transactions in the money market. A higher interest rate increases the liquidity of money. Given that the demand for money increases as national income increases, the relationship between i and Y is positive in the LM curve. Hence the IS–LM analysis leads to the determination of the interest rate and national income that describes the equilibrium between the product market (described by IS) and the money market (described by LM). As shown below, the environmental cost function can be considered alongside the IS–LM framework. Such consideration can lead to the need for choosing policies that change the IS–LM equilibrium to one that maintains the assimilative capacity of the environment.

Two possible contexts are presented in figures 13.1 and 13.2. The first situation in figure 13.1 represents the case where the IS–LM equilibrium results in a level of national income that is in excess of the limit set by the environmental cost function. The reverse of this situation is shown in figure 13.2. First, consider figure 13.1. The constraint set by the environmental cost function indicates that the level of national income has to be cut back from Y_o to Y_h. This can be achieved by one of three ways, as follows:

- The IS curve can be shifted inwards by raising taxes and cutting government spending, so that the IS–LM equilibrium moves from point **a** to point **b**. One possible argument is that if all environmental externalities are internalised in the product markets, say through the polluter pays principle, then the 'social IS curve' (if such a phrase can be used at all) is in fact the curve IS_h. The result is a lower interest rate and the level of output Y_h, which is in line with the environment. Given that the money stocks are kept constant, the fiscal intervention to comply with environment can be described as a 'tight fiscal–loose monetary' policy.
- The LM curve can be shifted to the left by cutting back on monetary supply so that an equilibrium in line with the environment, namely that at point **c**, is attained. This would represent a 'tight monetary–loose fiscal' policy.
- Both fiscal and monetary policies can be tightened so that an equilibrium such as that at point **d** is attained without impacting too much on interest rates.

Hence in figure 13.1 (p. 166), the economy is compelled to choose a set of contractionary policies. The reverse of this is shown in figure 13.2 (p. 166), where the economy can choose a set of expansionary policies. Consider now the effect on interest rates when the environment is included in IS–LM analysis. This effect will depend on whether the environment prompts an expansionary policy or a contractionary policy. For example, in the context of an expansionary policy, a 'tight monetary–loose fiscal' policy will result in a higher interest rate, while a 'tight fiscal–loose monetary' policy will result in a low interest rate. These policies have the reverse effect in the context of a contractionary policy.

However, some authors, for example Pearce and Turner (1990), indicate that the role of interest rates in sustaining the environment is unclear. The reasons for the ambiguity about interest rates are straightforward. A low interest rate extends the planning horizon, and society will prefer activities that generate the flow of goods and services over a long time period. This involves a shift from current consumption to investment. While the sustainable services of the environment are more likely to be recognised in the context of a low interest rate rather than a high interest rate, the act of investment is not without environmental costs. That is, the creation of investment goods also requires environmental inputs, and therefore it is not possible to conclude that low interest rates will sustain the environment. This is the root cause of the ambiguity.

However, the analysis within the IS–LM framework deals with nominal interest rates. Hence, the implications of the fiscal–monetary policy mix for changes in the price level become important. Generally a loose monetary policy can lead to an increase in the price level. Thus in figures 13.1 and 13.2, the difference between the equilibria at points **c** and **b** may not be significant in terms of real interest rates. However, the choice of policies for a sustainable environment is perhaps better served by constraining the process of output formation by an environmental cost

Interest rate

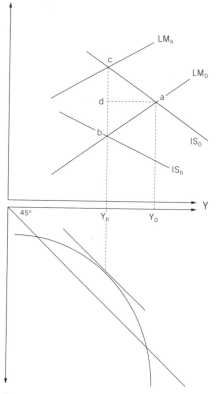

C_{EM}

Figure 13.1   Contractionary policies

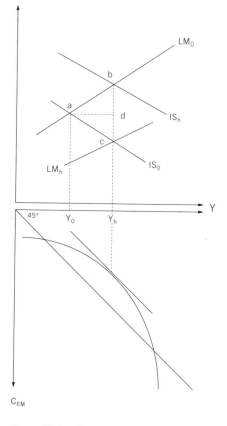

Figure 13.2   Expansionary policies

function as indicated in figures 13.1 and 13.2. That is, allow the interest rates to emerge from a set of policies that clearly recognise the limits imposed by the environment. The failure of either high or low interest rates to protect the environment is most likely due to the lack of information on environmental costs and the absence of such costs in the statement of national income accounts.

The environment and aggregate supply

The recognition of the environmental cost function can also alter the aggregate supply (AS) curve. In standard macroeconomic theory, the derivation of the aggregate supply curve combines the production function, the labour market, and the price level. A convenient way of introducing the environmental costs into the analysis of AS is to consider the impact of the environmental costs on the production function. This is shown in figure 13.3A, where $\{Y = Y(L)\}$ describes production in terms of labour (L) with fixed levels of capital (K) and technology (T), and ignores the effect of the environment, as in most standard analyses. $\{Y = Y_e(L)\}$ recognises the effects of the environmental depreciation. If L_F represents full employment, it is evident that the recognition of the environment leads to a reduction in employment.

Figure 13.3 (p. 168) is taken from Glahe (1977) and describes the derivation of the AS curve. As shown, this derivation of the AS curve (figure 13.3D) combines the production function (figure 13.3A), the labour market (figure 13.3B), and the relationship between price level (P) and real wage (w) (figure 13.3C). Along each schedule in figure 13.3C, the nominal wage is constant at the level specified, namely W_0, W_F, and W_2.

Initially, disregard the production function $\{Y = Y_e(L)\}$ and the labour demand and supply curves D_e and S_e. Also assume that the labour market is initially at an equilibrium with real wage w_F and full employment L_F with output Y_F being produced. If the price level at this equilibrium is P_F, then the nominal wage will be W_F. In figure 13.3D, (Y_F, P_F) is represented by point A on the aggregate supply curve labelled AS_F. If the labour market is imperfect with downwardly rigid nominal wages, then the response to changes in the lowering of the price level illustrates the segment BA of the aggregate supply curve AS_F. The upward flexibility of nominal wages when the price level increases (in the context of full employment) results in the vertical segment AE of this curve. Should nominal wages be flexible in both directions, then the aggregate supply curve will be a vertical line at Y_F.

Consider now the effect of introducing the environmental depreciation costs. The economy is now compelled to move to the lower production function $\{Y = Y_e(L)\}$. The level of national income falls from Y_F to Y_h and, due to the reduced

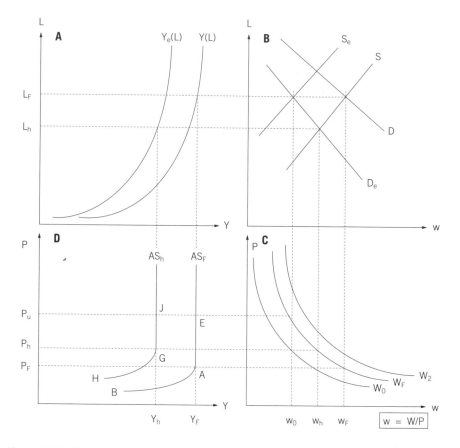

Figure 13.3   Environmental costs and aggregate supply

number of workers required, labour demand falls from D to D_e. Hence, real wages in the labour market will become w_h. Because nominal wages are downwardly rigid, there is pressure for the price level to rise from P_F to P_h. This is illustrated by point G on the aggregate supply curve, which is derived by recognising the environmental costs, namely AS_h.

When the price level falls from P_h, due to the downward rigidity of nominal wages, the segment HG of the curve AS_h will emerge. Hence, the recognition of the environment leads to a leftward shift in the AS curve, which becomes vertical at a lower level of output due to the environmental limit, and a higher price level when nominal wages are downwardly rigid. Flexibility of wages will result in the AS curve being a vertical line at Y_h.

Consider now the implications of the AS analysis. The aggregate supply curve AS_h implies an unemployment level of $(L_F - L_h)$ in the labour market. In terms of achieving full employment and general equilibrium, there are two options, as follows:

- Make investments in technologies that will restore the environment and render it more efficient so that the production function $Y_e(L)$ will shift towards $Y(L)$. This will result in the aggregate supply curve AS_h shifting to the right towards AS_F.
- Making prompt changes in the labour market by shifting (increasing) labour supply from S to S_e, so that a new lower real wage, w_0, with full employment emerges. If workers will not accept a lower nominal wage—that is, if they remain at W_F—the price level will be pressured to rise to P_u. This corresponds to point J on the curve AS_h. Alternatively, if they are willing to accept a lower nominal wage, say W_0, then the price level will be pressured to fall.

Hence, at least conceptually, the inflationary impact of recognising the environmental limits will be influenced by the extent of flexibility in the labour market. The price level changes that were suggested above are of course based on the premise that wage demands are the major causes of such changes, and the other causes are held constant. Similar conclusions are reached from adapting a Keynesian income–expenditure equilibrium model, which is considered next.

The environment and a general equilibrium model

An adapted Keynesian expenditure analysis is presented in figure 13.4 (p 170). This combines the standard expenditure (E)–income (Y) model in figure 13.4A with the production function (figure 13.4B), the aggregate supply (AS)–demand (AD) relationship (figure 13.4C), and the labour market (figure 13.4D). As in the previous section the effect of the environmental costs are introduced through the production function. Also suppose that the AS curves have been derived along the same lines as discussed above. Further, for convenience assume that nominal wages are flexible in both directions. Hence, the AS curves are vertical lines at the appropriate levels of output. The AD curve in figure 13.4C shows the equilibrium values of national income at different price levels, which in turn are associated with the different levels of expenditure that are shown in figure 13.4A. That is, for example, the price level associated with the schedule of expenditure E_0 is P_0, and the price level associated with the schedule E_F is P_F, and $(P_0 > P_F)$.

Suppose that the economy initially displays an equilibrium between E and Y (figure 13.4A) in terms of the national expenditure schedule labelled E_0. The level of production (Y_0) and real wage (w_0) are low, while the level of unemployment, namely $(L_F - L_0)$ is high. In order to achieve full employment, the Keynesian strategy will be to pump-prime the economy, so that the expenditure schedule shifts from E_0 to E_F to achieve full employment level of production, namely Y_F.

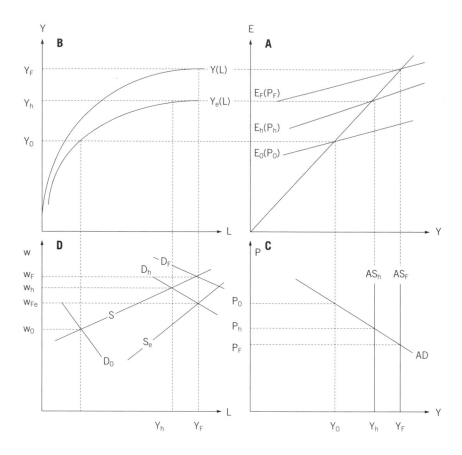

Figure 13.4 Adapted equilibrium analysis

The stimulus offered to the economy also raises the demand for labour from D_0 towards D_F with associated increases in real wages. However, due to the environmental limits that have been introduced through the production function, the expenditure schedule cannot be shifted beyond E_h. Hence the economy has to limit its output to Y_h, and the labour market will reveal a real wage of w_h with unemployment equalling $(L_F - L_h)$.

The inability to alleviate the unemployment problem by pump-priming an economy is usually ascribed to structural rigidities and market imperfections in the economy. For example when labour unions set barriers to entry and the log of wage claims is prohibitive, the usual Keynesian incentives will not have the desired employment and output effects. However, the constraints on the realisation of economy-wide outcomes are not confined to market and structural imperfections but include the environment as well. For example, if interest rates are lowered with the hope of stimulating the housing industry, and if environmental services to support such expansion have already been exhausted, then the incentives will have little effect.

As in the previous case, the options for alleviating the unemployment problem are to either:

- shift the production function upwards by investing in technologies that restore the environment and render it more efficient, or
- opt for lower real wages in the labour market; in other words, shift the labour supply outwards to S_e.

Note that the first option attempts to achieve the labour market equilibrium (L_F, w_F) and the AS–AD equilibrium (Y_F, P_F). That is, shifting the production upwards to raise the demand for labour results in shifting the AS curve from AS_h to AS_F. The second policy option is directed to the labour market equilibrium (L_F, w_{Fe}) and the AS–AD equilibrium (Y_h, P_h).

It is apparent, therefore, that a wages policy option to bring the economy in line with the environment could lead to a higher price level than the technology policy option. The environmental technology option can include innovations in fields such as molecular biology, biotechnology, and environmental engineering; for example a cost-effective biotechnology technique to treat algal blooms or a technique in chemical engineering to treat sewage. Such innovations are capable of shifting the environmental cost functions (figures 11.1 and 11.2) outwards, and thereby allowing output and demand to either expand or be maintained.

Concluding remarks

The object of this chapter has been to illustrate the effects of including the environment in selected macroeconomic frameworks. As observed above, the policy instruments are no doubt altered to bring the economy into line with the environment. A summary of the comparisons of the policy directives when the environment is included and when it is excluded from macroeconomic frameworks, is given in table 13.1 below.

Note that the low real wages policy option can mean that the high-income economies need to cut back on their growth and their lifestyles and the low-income countries need to set modest aspirations for their improvements in standards of living. As an example, consider some of the wasteful and extravagant consumption items in high-income countries. Most people of high-income countries drive to work—even drive to the nearby suburban shop that is a few metres away—mow their lawns and trim their edges with power tools, brush their teeth with electric toothbrushes, and carve their meat with electric carving knives. By the same token, low-income countries seem to be convinced that development means multistorey, billion-dollar hotel complexes along their coastlines, six-lane freeways, nuclear power stations, their own airlines with supersonic aircraft, and virtually everything the high-income countries have achieved to date.

So while the rich countries need to moderate their lifestyles, the poorer countries need to settle for less. Settling for less could mean such measures as power generation through windmills and solar panels, and the construction of buildings with locally available and less energy-intensive materials. In the meantime, mainstream economists need to change their attitudes too. For example, when an economy has to be revived out of a recession, economists need to distract themselves from the usual mechanisms of pump-priming the economy and consider incentives for investment in technologies that would render the natural environment more effective.

Table 13.1   Comparison of policy options

Macroeconomic framework environment	*Policy directive without environment*	*Policy directive with environment*
H–D model	Maximise the rate of economic growth by maximising savings ratio and minimising the capital–output ratio	Limit the rate of economic growth to the level set by the environmental cost function
IS–LM	Choose fiscal and monetary policies to maximise output with equilibrium in product and money markets	Choose fiscal and monetary policies to contract or expand output to the limits set by the environmental cost function
AS analysis and Keynesian model	Increase output to achieve full employment and general equilibrium	Achieve full employment and general equilibrium by investing in environmental technologies or choosing a lower real wage policy

REVIEW QUESTIONS

1 Illustrate with reference to the Harrod–Domar and IS–LM models the constraints on investment activities when environmental depreciation is internalised into these models.

2 Explain, by recourse to the derivation of aggregate supply, how the internalisation of environmental depreciation into the production function affects the determination of productive capacity.

3 Discuss the policies that an economic planner needs to consider in resolving the conflicts between maximising employment and maintaining environmental quality.

Part 4

The Scarcity Debate

14

Revisiting the Scarcity Debate

We commenced this text with a discussion of the scarcity debate and the evidence presented by the optimists and the pessimists. The major consensus of the optimists was that the scarcity of natural resources is a non-issue due to the dominance of technological change. In this chapter, we return to this debate and test the hypothesis of natural resource scarcity by developing a proxy method to value the stock of environmental capital that has been engaged in the formation of national product.

Environmental capital in aggregate output

In the established theory of capital pricing (Jorgenson 1967), the price of capital is made up of two components, namely the opportunity cost of money (that is, the interest rate) and the depreciation of capital stock. We estimate these two components for environmental capital and compare them with the prices of manufactured capital. The analysis centres on an aggregate Cobb–Douglas (C–D) utilisation function as proposed by Solow (1986). In this function, aggregate output (Y) is distributed between three factors, namely manufactured capital (KM), labour (L), and environmental capital (KN). That is:

$$Y = \alpha(KM)^{\theta}(L)^{\lambda}(KN)^{\eta} \qquad (14.1)$$

where θ, λ, and η respectively are shares of Y accruing to KM, L, and KN and $(\theta + \lambda + \eta)) = 1$ in the context of constant returns to scale and competitive markets.

This function is an extension of that used in most standard economic analyses where Y is assumed to be distributed between only KM and L (for example see Dornbusch & Fischer 1994). The central concept adopted here is that KN has a value in terms of utilisation for output only when economies deviate from the condition of perfect competition, because KN would be perfectly sustainable in the context of perfect competition. This conceptual premise is elaborated on below. The use of a C–D utilisation function for valuation purposes may appear difficult for some economists. This is because the C–D function explains a physical production technology that underlies output rather than the value of output. Nevertheless, the coefficients associated with the factors of utilisation, namely θ, λ, and η, explain the distribution of output. In this chapter, it is this distribution that is used as a basis for measurement and valuation. Further, as indicated below, though this is arduous, it becomes necessary to assume that the marginal value product that is derived from the C–D function can be approximated to a linear function.

The size of KN estimated here does not represent a nation's entire stock of natural endowments. Instead we conceptualise KN as a small subset that is drawn annually from the total set of endowments towards the formation of Y. Hence in equation (14.1), the distinction between KM and KN is that KM represents the size of accumulated stock while KN is the size of an annual stock that is regularly withdrawn from a much larger stock of unknown size. KN could include the waste receptacle facilities of air, water, and soil, forests, and other natural resources.

We also illustrate that the aggregate level valuation of KN requires nothing more than the standard macroeconomic data contained in the statements of national accounts including information on the magnitudes of KM, L, and wages.

The chapter is structured as follows. The conceptual and methodological premises are outlined in the next section. This is followed by the empirical illustration with respect to the data from Australia, Germany, and Sweden. The choice of Australia, Germany, and Sweden to illustrate this methodology was partially influenced by the availability of detailed historical information, particularly on the stocks of KM. The empirical analysis permits the quantification of relevant indicators for resource scarcity and efficiency. The policy implications of the time trends displayed by these indicators are then considered.

Perfect competition and perfect sustainability

As a basis for our analysis we offer the following proposition:

> *A state of perfect sustainability can prevail in the context of a perfectly competitive equilibrium and perfect sustainability could coexist only with a perfectly competitive equilibrium.*

By 'perfect sustainability' we mean the replenishment of KN to its natural level following its withdrawal. One could argue that the unique conditions of perfect competition (anonymity, homogeneity, perfect information, perfect mobility, and full employment) are such that it is possible for the imbalances imposed on nature to be restored through natural processes. That is: just as the state of perfect competition is unique, so is the state of perfect sustainability. For example the anonymity condition is such that any contaminating emission could be small enough not to exceed the assimilative capacity of nature. At the same time, if the condition of perfect information encompasses nature as well as the knowledge of the requirements to maintain the flow of services from nature, then it is possible to argue that individuals would voluntarily control their emissions. Though it would be tenuous, the argument could be extended to the case of non-renewable KN as well. That is, while anonymity could ensure that the magnitudes of withdrawals were small, perfect information (about user costs) could prompt the choice of renewable KN over non-renewable KN.

In just the same way as perfect competition is a theoretical construct that provides a basis for economic analyses, it is possible to argue that perfect sustainability too is a theoretical construct that should provide a basis for the analytics of environmental economics. If KN is perfectly sustainable, then its withdrawal could not entail an opportunity cost. As an example, consider the case of a small wine-maker who is based on a river bank. The wine-maker draws water from the river and then discharges the residues of the wine-making process into the river. If the river is capable of assimilating the residues and retaining its ecosystem properties so that the wine-maker does not have to treat the residues, then the wine-maker does not incur an opportunity cost for using the river. The river displays a state of perfect sustainability. Alternatively, consider a situation where the residues have to be treated prior to their discharge into the river because the river is unable to assimilate the residues without treatment. In this case the river may retain its properties and its sustainability, but has imposed an opportunity cost on the wine-maker. The distinction then between perfect sustainability and 'sustainability that is not perfect' lies in the presence or absence of opportunity costs. Hence, in the context of perfect competition, KN has zero opportunity cost due to its perfect sustainability. At the same time, however, the opportunity cost would be at its highest for both KM and L due to their full employment in the context of perfect competition.

The price of natural capital

Given the above conceptual premise, it is possible to estimate a value for KN in terms of the departure from the conditions of perfect competition. For reasons of clarity we present our case in point form below.

1   In the context of perfect competition, Y is completely distributed between KM and L because KN has no user cost value due to its sustainability. That is,

in equation (14.1), $\eta = 0$ and $\{(\theta + \lambda) = 1\}$. This context is illustrated in figure 14.1, which describes an aggregate market for KM and is an adaptation of the exposition by Samuelson (1970). We assume that the marginal value product function for KM from equation (14.1) can be approximated to an aggregate linear demand function $\{D_{KM} = f(KM)\}$ and also that the marginal cost of KM is constant at P_{KM}. With perfect competition the following definitions hold:

$$Y = \int_{0}^{KM_1} (f(KM)).\, dKM, \tag{14.2}$$

$$OS = P_{KM} * (KM_1) \tag{14.3}$$

$$SW = Y - OS \tag{14.4}$$

where KM_1 represents the utilisation level of KM, while OS is the sum of all payments to KM, namely the operating surplus, and SW is the sum of all payments to L, namely the sum of wages.

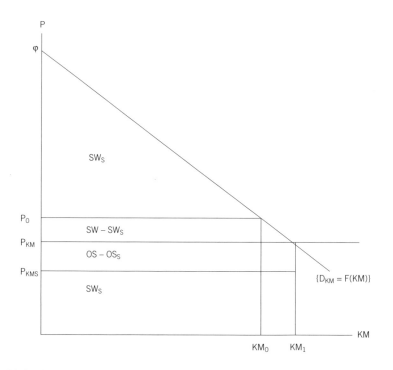

Figure 14.1   The aggregate market for KM and the distribution of Y

2 Consider now the case of imperfect competition. KN is no longer sustainable and therefore has user cost value. This value can be illustrated in terms of the deviation that has emerged from the definitions presented for the case of perfect competition.

When the labour market is imperfect, SW will no longer be a true estimate of the opportunity cost of L and has to be reduced to its shadow value (SW_S) by an appropriate amount. In figure 14.1, SW_S is defined as:

$$SW_S = \int_0^{KM_0} (f(KM)) \, dKM - (P_0 * KM_0) \tag{14.5}$$

When the KM market is imperfect, the OS will no longer be a true estimate of the opportunity cost of KM and this too has to be reduced to its shadow value (OS_S). In figure 14.1, this revised value is defined as:

$$OS_S = P_{KMS} * KM_1 \tag{14.6}$$

3 Hence, with imperfect competition, there is a proportion of Y that is unaccounted for in terms of true opportunity costs. Since we assume in equation (14.1) that Y describes the utilisation of only three factors (KM, L, and KN),[5] this proportion has to be allocated to KN and is defined as the sum of the respective amounts by which SW and OS need to be reduced. That is:

$$(\eta*Y) = (SW - SW_S) + (OS - OS_S) \tag{14.7}$$

From figure 14.1, it is apparent that the price differential ($P_0 - P_{KMS}$) becomes an indicator of the user cost value of KN. In other words, the greater the size of this differential, the greater the deviation from perfect competition and accordingly the greater the unsustainability of KN and its user cost value. Because a zero price for KN corresponds to a zero value for the differential ($P_0 - P_{KMS}$), it is possible to nominate ($P_0 - P_{KMS}$) as the price of KN, namely P_{KN}. Hence it follows that the size of KN that is used for Y is:

$$KN = (\eta*Y) / (P_0 - P_{KMS}) \tag{14.8}$$

4 Following the definitions of OS_S and SW_S above, the coefficients in equation (14.1) that describe the factor shares of Y in the context of imperfect competition can be defined as:

$$\theta = (OS_S)/Y \tag{14.9}$$

$$\lambda = (SW_S)/Y \tag{14.10}$$

$$\eta = 1 - \theta - \lambda \tag{14.11}$$

5   The assumption of constant marginal costs enables the prices of KM to be equated to average cost and hence, from (14.3) and (14.6) above, the prices P_{KMS} and P_{KM} can be defined as:

$$P_{KMS} = (OS_S)/KM_1 \tag{14.12}$$

$$P_{KM} = (OS)/KM_1 \tag{14.13}$$

Given the assumption of linearity for $\{D = f(KM)\}$, supposing $\{f(KM) = \varphi - \gamma KM\}$, P_0 can be defined as follows:

$$P_0 = \varphi - \{(2^*\gamma^*SW_S)\}^{(1/2)} \tag{14.14}$$

where:

$$\gamma = (\varphi - P_{KM})/KM_1 \tag{14.15}$$

$$\varphi = (2SW/KM_1) + P_{KM} \tag{14.16}$$

Hence the price of KN can be estimated as:

$$P_{KN} = [\varphi - \{(2^*\gamma^*SW_S)\}^{(1/2)}] - \{(OS_S)/KM_1\} \tag{14.17}$$

6   As indicated, the price of capital consists of two components:
   a   the interest rate (or the opportunity cost of money), namely the sacrifice made of present consumption for future consumption
   b   the depreciation rate, which is the amount of replacement capital that is needed for every unit of additional capital that is accumulated.

Because the information on capital consumption (replacement) for KM is readily available, it is possible to estimate the depreciation rate for KM (δ_{KM}) directly from the national accounts:

$$\delta_{KM} = (KM_C) / KM_1 \tag{14.18}$$

where KM_C is the quantity of capital consumed during the period when the accumulated stock equals KM_1.

Since both P_{KMS} and δ_{KM} can be directly estimated from the national accounts—equations (14.12) and (14.18)—it is possible to estimate the interest rate for KM (i_{KM}) as a rate that is implied from the national accounts by recourse to the Jorgenson (1967) formula for the price of capital:

$$i_{KM} = (P_{KMS}/D) - \delta_{KM} \tag{14.19}$$

where D is a capital stock deflator.

7   The interest rate for KN (i_{KN}) can be estimated on the basis of equivalence between KN and KM. For example, if during a given time period the quantities of KN and KM used are respectively KN_1 and KM_1, then:

$$i_{KN} = i_{KM} * (KN_1 / KM_1) \qquad (14.20)$$

This estimate is based on the premise that one unit of KN is equivalent to (KN_1/KM_1) units of KM and hence the opportunity cost of money for one unit of KN is apportioned on the basis of this equivalence.

The Jorgenson (1967) formula can now be extended to the case of KN to estimate the depreciation rate (δ_{KN}) by combining equations (14.17) and (14.20). That is:

$$\delta_{KN} = (P_{KN}/D) - i_{KN} \qquad (14.21)$$

The estimation of the depreciation rate, δ_{KN}, in turn permits the estimation of the magnitude of environmental capital consumption (KN_C), a quantity that is analogous to KM_C. This can be defined by combining equations (14.8) and (14.21):

$$KN_C = (\delta_{KN}) * (KN) \qquad (14.22)$$

Hence sustainable income can be defined as the difference $(Y - KN_C)$, and the ratio (KN_C/Y) becomes an important indicator of the scarcity of KN.

We next illustrate the conceptual definitions given above in the context of Germany, Sweden, and Australia.

Empirical illustration

We adopt a relatively simple approach to estimate SW_S by assuming that the dominant source of imperfection in the labour market is unemployment. We estimate the shadow wage rate (W_S) through dividing SW by the labour force (L_F)—namely those employed as well as those unemployed—and then SW_S is determined as the product of W_S and the actual level of employment (L_A). That is:

$$(SW_S) = (W_S) * (L_A) \qquad (14.23)$$

Hence in figure 14.1, the reduction in SW that is required to reflect competitiveness is defined by:

$$(SW) - (SW_S) = \{L_A * (W - W_S)\} \qquad (14.24)$$

Where W is market wage, namely (SW/L_A)

The association between this simple approach to the shadow pricing of labour and the sustainability of KN can be illustrated as follows. In the context of unemployment a certain proportion of society receives very low (or zero) income while the rest of society receives a disproportionately higher income. There is indeed empirical evidence from developing countries (Anderson & Thampapillai 1990) to confirm that poverty and unemployment prompt the degradation of KN. Such evidence for developed countries is anecdotal following the observation of higher than usual depletion of natural resources during periods of high unemployment. At the same time, the disproportionately higher incomes that are afforded to those in employment could prompt extravagant patterns of consumption, which in turn lead towards the degradation of KN. Therefore, the lowering of wages as suggested in (14.23) and (14.24) above could increase employment and at the same time constrain extravagant patterns of consumption, both of which could in turn help reduce the degradation of KN.

A relatively straightforward approach is also adopted to estimate OS_S. We assume that the dominant source of imperfection in the market for KM is the presence of various taxes and subsidies that surround the utilisation of KM. Because in the context of circular flow, subsidies are in fact a proportion of taxes that are returned to the owners of capital, OS_S is defined as the difference between OS and net taxes (NT), which are taxes less subsidies:

$$OS_S \ = \ OS - NT \tag{14.25}$$

In reality, taxes strive to correct divergences between private and social values due not only to the unsustainability of KN but also to other sources of imperfection. However, a perfectly competitive equilibrium that could ensure the state of perfect sustainability is also without taxes and any form of transfer payments. It is possible, therefore, to argue that a shadow pricing approach based on the removal of taxes (transfer payments) would eventually lead to a competitive equilibrium in the long run.

For reasons of consistency across the various countries, we have used Net Domestic Product as the measure of Y, and the data were taken from the National Accounts compiled by the OECD (1997, 1998, and 1999). The data for Germany refer to the former West Germany, and the series is not extended beyond 1994 due to changes in the data-reporting system following reunification in 1989.

Tables 14.1 and 14.2 provide the estimates of market and shadow values for factor incomes and prices of KM and L. The distribution of Y between the three factors, after KM and L have been shadow priced, is shown in table 14.3. It is evident that the share of Y accruing to KN, namely η, has been increasing in all three economies. This is indicative of the fact that the economies have continued to deviate from the benchmark of PC. While Germany reveals a much lower value for η than do the other two economies, the rate of increase of η in Germany has been much higher. Table 14.4 illustrates the price and quantity of KN, including the size of capital consumption (KN_C). For reasons of clarity all tables are placed at the end of the chapter.

The scarcity debate

The estimates derived here can be discussed under two headings:
- the relative scarcity of KN
- the efficiency with which KN is utilised in the economy.

We consider each of these in turn below.

The relative scarcity of KN

The scarcity debate of the 1970s (Forrester 1971; Meadows et al. 1972) invoked the analysis of natural resource prices. The main hypothesis in these analyses was the contention that if natural resources are becoming scarce (due to progressively higher rates of extraction and depletion), then their relative prices must rise. Most analyses had rejected the hypothesis on the grounds of falling natural resource prices (Nordhaus 1973; World Bank 1992). For example, as indicated in chapter 1, the World Bank (1992, p. 37), by recourse to the analysis of non-ferrous metal prices between 1900 and 1991, reported: 'Declining price trends also indicate that many non-renewables have become more, rather than less, abundant.'

The observed falls in price are attributed to technological improvements in resource utilisation (Samuelson & Nordhaus 1990; Mankiw 1998). The analysis here does not fully support this. For example in figure 14.2, where trend lines have been fitted for the point estimates of P_{KN} over a 28-year period (1969–97), we observe an initial declining trend in both Germany and Sweden followed, however,

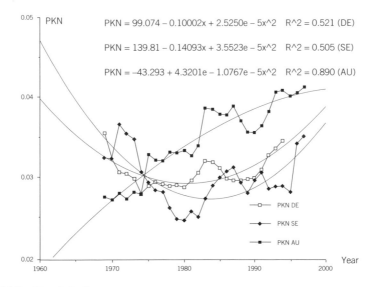

$$PKN = 99.074 - 0.10002x + 2.5250e - 5x^2 \quad R^2 = 0.521 \text{ (DE)}$$

$$PKN = 139.81 - 0.14093x + 3.5523e - 5x^2 \quad R^2 = 0.505 \text{ (SE)}$$

$$PKN = -43.293 + 4.3201e - 1.0767e - 5x^2 \quad R^2 = 0.890 \text{ (AU)}$$

Figure 14.2 Trends in P_{KN}

by an increasing trend. Australia displays an increasing trend throughout the entire period. Two other scarcity indicators, the ratio of relative prices (P_{KN}/P_{KM}) and the ratio of environmental depreciation to NDP (KN_C/Y), are presented in table 14.5. These too do not unequivocally endorse a declining trend. It is possible to argue that the declining price trends are likely to be less dramatic when prices are estimated in terms of true opportunity costs.

The efficiency of utilisation of KN

One of the important arguments in the resource scarcity debate has been the concurrence of opinion that improvements in technology would offset the constraints imposed by scarcity. The ratio (KN/KM), which measures the quantity of KN required for the formation of one unit of KM, is an indicator of the efficiency with KN. This ratio remains almost constant for Australia and Sweden, and displays a marginally increasing trend for Germany (table 14.5). It is possible to argue that the technological advances with respect to the utilisation of KN are overstated when measured outside the calculus of shadow prices. This observation is tested further by examining point estimates of two more ratios, namely ($\Delta KM/Y$) and (KN/ΔKM).

($\Delta KM/Y$) measures the additional amount of KM that is needed each year for the formation of one unit of Y. As illustrated in table 14.6 and figure 14.3, this ratio displays a declining trend, reinforcing the commonly held belief that techno-

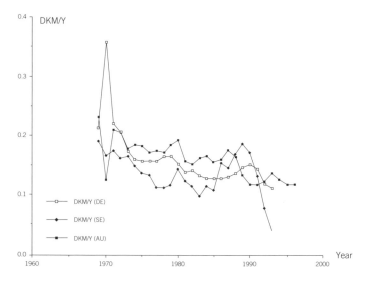

Figure 14.3   Trends in ΔKM/Y

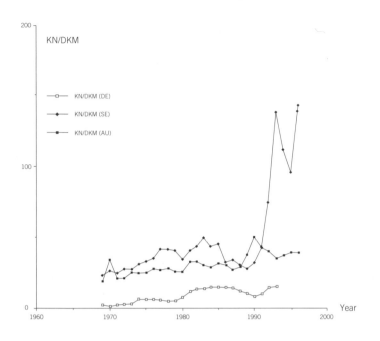

Figure 14.4 Trends in KN/ΔKM

logical advances and efficiency gains have been significant in the utilisation of KM. However, as shown in figure 14.4, the ratio (KN/ΔKM) moves in a direction opposite to the ratio (ΔKM/Y). Note that (KN/ΔKM) is the amount of KN that is needed for the formation of one additional unit of KM. The increasing trend of (KN/ΔKM) in conjunction with the decreasing trend of (ΔKM/Y) indicates that, while economies have gained momentum with respect to the efficiency of utilising KM, they have been increasingly inefficient in the utilisation of KN.

Concluding remarks

We have presented here a simple alternative framework for the valuation of KN at the aggregate level. As indicated, these values can be derived as point estimates directly from the national accounts with the aid of some simplifying assumptions. This is clearly advantageous in the context of sparse data for environmental variables. At the same time, the analysis provides useful information for economic planning at the national level in at least two ways. First, it has been possible to determine the proportion of national income that has to be set aside for the sustainability of KN. This is the depreciation allowance for KN and has been defined above as KN_C. For the late 1990s, this proportion has been in the order of 5 to 8 per cent in Germany and Australia, and has exceeded 10 per cent in Sweden. When economic planners fail to recognise this, they are in fact overstating the performance of the economy by an amount equivalent to the depreciation allowance. An appreciation of the magnitude of this allowance enables the planner to revise targets and instruments.

Second, the analysis of scarcity and efficiency ratios reveals that much effort needs to be expended in the area of environmental capital efficiency. As indicated, although efficiency gains have been very significant in the utilisation of KM, these have been lacking with respect to KN. Therefore, there is an explicit need to search for technical improvements that directly target KN. These could include the development of cost-effective methods of waste management and a transition to the utilisation of renewable KN from non-renewable KN. Further, this also implies that during periods of stagnation the enhancement of KN could become a vehicle for economic recovery. Recall that this inference was drawn in chapters 12 and 13 as well.

It is pertinent to conclude this text with the words of Alfred Marshall (1891):

Man does not create things. He only rearranges matter.

This is in no uncertain terms the first law of thermodynamics. When matter is rearranged, the second law of thermodynamics takes effect. It is therefore difficult to ignore the issue of environmental scarcity, because economic activity inevitably involves the rearrangement of KN and an increase in its entropic value and hence scarcity. This result was illustrated in this chapter using the standard tools of analysis.

Table 14.1 Factor incomes for KM and L—market and shadow values

Year	OS (DE) DM (10^6)	OS_S (DE) DM (10^6)	SW (DE) DM (10^6)	SW_S (DE) DM (10^6)	OS (SE) SEK (10^6)	OS_S (SE) SEK (10^6)	SW (SE) SEK (10^6)	SW_S (SE) SEK (10^6)	OS (AU) $A (10^6)	OS_S (AU) $A (10^6)	SW (AU) $A (10^6)	SW_S (AU) $A (10^6)
1969	559 391	377 784	716 905	701 684	244 111	148 558	524 242	513 909	69 507	51 115	101 269	98 914
1970	545 461	375 496	790 300	773 520	260 436	160 700	558 016	547 213	68 630	49 594	111 189	108 603
1971	540 768	365 011	830 642	813 005	257 815	138 617	569 775	558 773	72 174	51 894	116 594	113 883
1972	554 180	371 153	873 565	855 018	267 045	147 085	582 967	571 711	77 830	57 055	119 220	116 447
1973	562 582	375 893	934 410	914 570	290 003	167 415	585 946	574 690	78 305	55 863	127 066	124 111
1974	527 032	346 253	964 679	944 196	280 202	169 970	619 102	606 698	68 789	45 753	137 867	134 661
1975	512 642	335 574	952 678	915 039	277 693	166 563	647 468	636 961	73 675	47 814	138 054	131 830
1976	554 785	368 519	991 955	953 060	250 707	140 722	686 278	675 212	77 301	51 112	140 162	133 602
1977	563 472	371 750	1 027 507	988 365	218 875	107 573	698 196	685 651	75 114	49 388	142 862	134 857
1978	585 819	387 594	1 052 151	1 013 757	222 928	121 037	709 632	693 783	87 603	60 084	141 678	132 946
1979	608 165	397 583	1 096 769	1 061 868	242 878	143 892	722 546	707 648	92 079	62 748	142 047	133 841
1980	575 531	360 683	1 135 966	1 099 832	252 589	150 549	726 841	712 365	91 889	61 373	150 037	141 255
1981	559 627	345 340	1 143 257	1 091 881	250 942	143 791	725 719	707 627	89 683	58 216	157 069	148 329
1982	551 405	342 280	1 127 612	1 055 236	274 092	171 450	708 922	686 631	82 612	50 206	157 744	147 224
1983	592 700	378 129	1 114 769	1 026 772	299 908	184 516	698 043	673 950	101 212	66 356	156 033	140 741

cont.

Table 14.1   Factor incomes for KM and L—market and shadow values (cont.)

Year	OS (DE) DM (10^6)	OS_S (DE) DM (10^6)	SW (DE) DM (10^6)	SW_S (DE) DM (10^6)	OS (SE) SEK (10^6)	OS_S (SE) SEK (10^6)	SW (SE) SEK (10^6)	SW_S (SE) SEK (10^6)	OS (AU) $A (10^6)	OS_S (AU) $A (10^6)	SW (AU) $A (10^6)	SW_S (AU) $A (10^6)
1984	622 438	403 998	1 132 123	1 042 609	331 774	202 862	709 924	687 936	108 139	70 177	163 048	149 257
1985	638 728	421 793	1 151 854	1 060 015	334 997	195 996	725 769	705 262	111 879	72 381	169 337	156 103
1986	663 588	450 618	1 174 295	1 084 658	342 968	194 947	744 402	724 709	114 161	73 360	172 090	158 568
1987	663 857	448 554	1 200 796	1 109 713	352 171	191 326	769 050	754 438	126 670	82 534	175 206	161 600
1988	707 047	485 806	1 229 764	1 136 598	354 694	197 314	788 631	775 931	136 536	91 457	179 155	167 026
1989	751 887	512 982	1 254 881	1 169 058	343 746	186 940	823 518	812 421	137 559	91 693	188 223	177 099
1990	807 470	554 080	1 315 520	1 233 952	329 761	159 667	851 189	837 128	132 055	87 413	190 386	176 738
1991	851 766	569 971	1 376 594	1 299 918	335 065	164 023	831 654	807 161	132 637	89 234	181 134	163 582
1992	851 798	555 075	1 409 971	1 316 475	335 444	191 613	814 254	771 234	139 316	95 186	195 319	174 219
1993	822 706	523 770	1 380 781	1 272 212	336 221	209 847	777 500	713 428	149 195	100 566	203 347	180 987
1994	883 293	569 818	1 375 522	1 260 455	374 579	245 134	784 984	722 450	158 716	105 681	212 763	192 173
1995					427 714	298 639	789 302	728 445	166 667	110 976	221 333	202 617
1996					408 811	243 196	827 687	761 050	167 178	109 990	232 964	213 063
1997					421 367	248 766	839 864	772 502	174 960	115 018	241 344	220 708

Table 14.2   Market and shadow prices for KM and L

Year	P_{KM} (DE) %	P_{KMS} (DE) %	$i_{(KM)}$ (DE) %	δ_{KM} (DE) %	W (DE) DM	W_s (DE) DM	P_{KM} (SE) %	P_{KMS} (SE) %	i_{KM} (SE) %	δ_{KM} (SE) %	W (SE) SEK	W_s (SE) SEK	P_{KM} (AU) %	P_{KMS} (AU) %	i_{KM} (AU) %	δ_{KM} (AU) %	W (AU) $A	W_s (AU) $A
1969	0.1009	0.0682	0.0433	0.0249	33 997	33 275	0.0748	0.0455	0.0205	0.0250	138 688	135 955	0.0921	0.0677	0.0345	0.0332	19 408	18 957
1970	0.0936	0.0645	0.0388	0.0257	36 225	35 456	0.0761	0.0470	0.0197	0.0273	144 939	142 133	0.0861	0.0622	0.0285	0.0338	20 777	20 293
1971	0.0856	0.0578	0.0325	0.0252	37 851	37 048	0.0724	0.0389	0.0120	0.0269	147 610	144 760	0.0879	0.0632	0.0284	0.0348	21 242	20 748
1972	0.0836	0.0560	0.0306	0.0254	39 620	38 779	0.0720	0.0396	0.0131	0.0266	151 028	148 112	0.0903	0.0662	0.0318	0.0343	21 177	20 685
1973	0.0810	0.0541	0.0289	0.0252	41 913	41 023	0.0752	0.0434	0.0166	0.0268	151 017	148 116	0.0865	0.0617	0.0282	0.0336	22 007	21 495
1974	0.0732	0.0481	0.0226	0.0255	42 695	41 788	0.0699	0.0424	0.0143	0.0281	156 260	153 129	0.0730	0.0486	0.0129	0.0356	23 280	22 739
1975	0.0689	0.0451	0.0197	0.0254	43 269	41 559	0.0670	0.0402	0.0128	0.0274	159 396	156 810	0.0751	0.0487	0.0129	0.0358	23 359	22 306
1976	0.0722	0.0480	0.0225	0.0255	45 362	43 583	0.0586	0.0329	0.0056	0.0273	167 876	165 169	0.0757	0.0500	0.0142	0.0359	23 490	22 390
1977	0.0710	0.0468	0.0215	0.0253	46 879	45 093	0.0497	0.0244	−0.0032	0.0277	170 333	167 273	0.0709	0.0466	0.0102	0.0365	23 555	22 235
1978	0.0715	0.0473	0.0220	0.0253	47 690	45 949	0.0495	0.0269	−0.0011	0.0280	172 450	168 599	0.0797	0.0546	0.0184	0.0363	23 321	21 884
1979	0.0717	0.0469	0.0211	0.0258	49 053	47 492	0.0526	0.0312	0.0027	0.0285	172 858	169 294	0.0808	0.0551	0.0189	0.0361	23 101	21 766
1980	0.0657	0.0411	0.0148	0.0263	49 921	48 333	0.0534	0.0318	0.0033	0.0285	171 749	168 328	0.0776	0.0518	0.0155	0.0363	23 617	22 234
1981	0.0620	0.0383	0.0115	0.0267	50 218	47 961	0.0515	0.0295	0.0012	0.0283	171 809	167 525	0.0728	0.0472	0.0110	0.0362	24 292	22 940
1982	0.0595	0.0370	0.0103	0.0267	50 018	46 808	0.0548	0.0343	0.0058	0.0285	167 991	162 709	0.0650	0.0395	0.0032	0.0363	24 449	22 818

cont.

Table 14.2 Market and shadow prices for KM and L (cont.)

Year	P_{KM} (DE) %	P_{KMS} (DE) %	$i_{(KM)}$ (DE) %	δ_{KM} (DE) %	W (DE) DM	W_S (DE) DM	P_{KM} (SE) %	P_{KMS} (SE) %	i_{KM} (SE) %	δ_{KM} (SE) %	W (SE) SEK	W_S (SE) SEK	P_{KM} (AU) %	P_{KMS} (AU) %	i_{KM} (AU) %	δ_{KM} (AU) %	W (AU) $A	W_S (AU) $A
1983	0.0623	0.0398	0.0132	0.0266	50 455	46 472	0.0586	0.0361	0.0073	0.0288	165 256	159 553	0.0773	0.0507	0.0152	0.0354	24 712	22 290
1984	0.0639	0.0415	0.0147	0.0267	51 171	47 125	0.0636	0.0389	0.0101	0.0287	166 845	161 677	0.0798	0.0518	0.0165	0.0353	24 942	22 833
1985	0.0640	0.0423	0.0157	0.0266	51 472	47 368	0.0627	0.0367	0.0079	0.0288	168 823	164 053	0.0798	0.0516	0.0146	0.0370	25 098	23 137
1986	0.0650	0.0441	0.0180	0.0261	51 456	47 529	0.0628	0.0357	0.0073	0.0284	174 374	169 761	0.0789	0.0507	0.0128	0.0380	24 623	22 688
1987	0.0635	0.0429	0.0170	0.0259	52 145	48 190	0.0625	0.0340	0.0055	0.0285	177 323	173 954	0.0847	0.0552	0.0179	0.0373	24 463	22 564
1988	0.0660	0.0454	0.0194	0.0260	53 026	49 009	0.0611	0.0340	0.0052	0.0288	179 275	176 388	0.0880	0.0590	0.0227	0.0362	24 135	22 501
1989	0.0684	0.0467	0.0204	0.0262	53 809	50 129	0.0573	0.0312	0.0021	0.0291	184 397	181 912	0.0857	0.0571	0.0211	0.0360	23 981	22 563
1990	0.0714	0.0490	0.0222	0.0268	55 077	51 662	0.0530	0.0257	-0.0031	0.0288	190 636	187 487	0.0801	0.0530	0.0177	0.0354	23 984	22 265
1991	0.0732	0.0490	0.0214	0.0275	44 323	41 854	0.0522	0.0255	-0.0022	0.0277	189 745	184 157	0.0786	0.0529	0.0181	0.0348	23 360	21 096
1992	0.0712	0.0464	0.0187	0.0277	46 246	43 179	0.0510	0.0291	0.0021	0.0270	194 101	183 846	0.0807	0.0551	0.0199	0.0352	25 245	22 518
1993	0.0672	0.0428	0.0151	0.0277	46 047	42 427	0.0504	0.0315	0.0041	0.0274	196 140	179 977	0.0843	0.0568	0.0215	0.0353	26 279	23 389
1994	0.0707	0.0456	0.0184	0.0273	46 512	42 621	0.0558	0.0365	0.0097	0.0268	199 843	183 923	0.0871	0.0580	0.0234	0.0346	26 662	24 082
1995							0.0631	0.0441	0.0182	0.0259	198 018	182 751	0.0890	0.0593	0.0252	0.0341	26 689	24 432
1996							0.0597	0.0355	0.0102	0.0253	208 854	192 039	0.0871	0.0573	0.0233	0.0340	27 790	25 416
1997							0.0611	0.0361	0.0109	0.0252	214 142	196 966	0.0888	0.0584	0.0244	0.0340	28 528	26 088

Table 14.3 Factor shares of income

Year	θ (DE)	λ (DE)	η (DE)	θ (SE)	λ (SE)	η (SE)	θ (AU)	λ (AU)	η (AU)
1969	0.296	0.689	0.015	0.1919	0.6638	0.1443	0.2993	0.5792	0.1215
1970	0.281	0.704	0.015	0.1948	0.6634	0.1418	0.2758	0.6040	0.1202
1971	0.266	0.718	0.016	0.1668	0.6723	0.1609	0.2749	0.6033	0.1218
1972	0.260	0.724	0.016	0.1731	0.6729	0.1539	0.2895	0.5910	0.1195
1973	0.251	0.733	0.016	0.1897	0.6513	0.1590	0.2720	0.6043	0.1237
1974	0.232	0.752	0.016	0.1878	0.6705	0.1417	0.2214	0.6516	0.1270
1975	0.229	0.741	0.030	0.1791	0.6850	0.1359	0.2258	0.6226	0.1515
1976	0.238	0.732	0.030	0.1501	0.7201	0.1298	0.2350	0.6144	0.1506
1977	0.234	0.737	0.029	0.1175	0.7487	0.1339	0.2266	0.6187	0.1547
1978	0.237	0.736	0.028	0.1302	0.7463	0.1235	0.2621	0.5798	0.1581
1979	0.233	0.742	0.024	0.1491	0.7334	0.1174	0.2680	0.5717	0.1603
1980	0.211	0.764	0.025	0.1537	0.7273	0.1190	0.2537	0.5839	0.1624
1981	0.203	0.761	0.036	0.1472	0.7245	0.1282	0.2359	0.6011	0.1629
1982	0.204	0.745	0.051	0.1744	0.6985	0.1271	0.2089	0.6125	0.1786
1983	0.221	0.717	0.061	0.1849	0.6753	0.1398	0.2579	0.5471	0.1949

cont.

Table 14.3   Factor shares of income (cont.)

Year	θ (DE)	λ (DE)	η (DE)	θ (SE)	λ (SE)	η (SE)	θ (AU)	λ (AU)	η (AU)
1984	0.230	0.709	0.061	0.1947	0.6604	0.1449	0.2588	0.5504	0.1908
1985	0.236	0.703	0.061	0.1848	0.6649	0.1504	0.2574	0.5551	0.1875
1986	0.245	0.697	0.058	0.1793	0.6665	0.1542	0.2563	0.5539	0.1898
1987	0.241	0.702	0.058	0.1706	0.6729	0.1565	0.2734	0.5353	0.1913
1988	0.251	0.692	0.057	0.1726	0.6787	0.1488	0.2897	0.5291	0.1812
1989	0.256	0.693	0.051	0.1602	0.6960	0.1438	0.2815	0.5436	0.1749
1990	0.261	0.693	0.046	0.1352	0.7089	0.1559	0.2711	0.5481	0.1808
1991	0.256	0.703	0.041	0.1406	0.6918	0.1676	0.2758	0.5055	0.2187
1992	0.245	0.705	0.050	0.1669	0.6719	0.1612	0.2844	0.5206	0.1949
1993	0.238	0.702	0.060	0.1884	0.6406	0.1710	0.2852	0.5132	0.2016
1994	0.252	0.685	0.063	0.2114	0.6230	0.1656	0.2845	0.5173	0.1982
1995				0.2454	0.5986	0.1561	0.2860	0.5222	0.1918
1996				0.1967	0.6155	0.1878	0.2749	0.5325	0.1927
1997				0.1972	0.6125	0.1903	0.2763	0.5302	0.1936

Table 14.4 Prices, stock size, and consumption quantities for KN

Year	P_{KN} (DE) %	i_{KN} (DE) %	δ_{KN} (DE) %	KN (DE) DM (10^9)	KN_C (DE) DM (10^9)	P_{KN} (SE) %	i_{KN} (SE) %	δ_{KN} (SE) %	KN (SE) SEK (10^9)	KN_C (SE) SEK(10^9)	P_{KN} AU %	i_{KN} AU %	δ_{KN} AU %	KN AU $A ($10^6$)	KN_C AU $A ($10^6$)
1969	0.0355	0.0042	0.0313	537 025	16 826	0.0324	0.0216	0.0108	3 444 205	37 268	0.0275	0.0345	-0.0070	754 217	-5 286
1970	0.0321	0.0042	0.0278	635 726	17 699	0.0323	0.0208	0.0115	3 620 663	41 684	0.0272	0.0284	-0.0013	796 143	-1 027
1971	0.0306	0.0036	0.0270	697 565	18 861	0.0366	0.0123	0.0243	3 656 687	88 686	0.0280	0.0284	-0.0004	820 124	-304
1972	0.0304	0.0034	0.0270	737 752	19 922	0.0354	0.0130	0.0224	3 697 802	82 688	0.0273	0.0318	-0.0045	861 713	-3 873
1973	0.0298	0.0033	0.0264	799 720	21 137	0.0347	0.0174	0.0173	4 040 743	70 059	0.0281	0.0281	0.0000	904 268	-45
1974	0.0280	0.0027	0.0252	869 946	21 947	0.0306	0.0150	0.0156	4 187 461	65 458	0.0279	0.0129	0.0150	941 461	14 079
1975	0.0289	0.0041	0.0248	1 543 607	38 319	0.0293	0.0133	0.0161	4 306 021	69 125	0.0328	0.0128	0.0199	979 262	19 505
1976	0.0294	0.0046	0.0247	1 573 580	38 942	0.0283	0.0056	0.0227	4 299 484	97 486	0.0321	0.0141	0.0180	1 018 927	18 349
1977	0.0291	0.0043	0.0248	1 593 805	39 553	0.0281	-0.0032	0.0313	4 355 643	136 462	0.0320	0.0101	0.0218	1 055 549	23 025
1978	0.0289	0.0042	0.0247	1 578 152	38 956	0.0261	-0.0011	0.0272	4 393 815	119 618	0.0331	0.0183	0.0148	1 095 378	16 222
1979	0.0290	0.0036	0.0254	1 435 285	36 474	0.0247	0.0027	0.0220	4 591 188	100 974	0.0330	0.0189	0.0142	1 136 141	16 080
1980	0.0287	0.0025	0.0261	1 499 006	39 174	0.0246	0.0033	0.0214	4 730 484	101 118	0.0333	0.0155	0.0178	1 180 816	21 016
1981	0.0295	0.0026	0.0269	2 068 496	55 553	0.0257	0.0012	0.0245	4 871 102	119 216	0.0327	0.0110	0.0217	1 228 571	26 682
1982	0.0305	0.0031	0.0274	2 811 371	77 026	0.0250	0.0058	0.0193	4 990 019	96 140	0.0339	0.0031	0.0308	1 265 313	38 941

cont.

Table 14.4 Prices, stock size, and consumption quantities for KN (cont.)

Year	P_{KN} (DE) %	i_{KN} (DE) %	δ_{KN} (DE) %	KN (DE) DM(10^6)	KN_C (DE) DM(10^6)	P_{KN} (SE) %	i_{KN} (SE) %	δ_{KN} (SE) %	KN (SE) SEK(10^6)	KN_C (SE) SEK(10^6)	P_{KN} AU %	i_{KN} AU %	δ_{KN} AU %	KN AU $A($10^6$)	KN_C AU $A($10^6$)
1983	0.0320	0.0046	0.0275	3 277 885	89 999	0.0273	0.0073	0.0201	5 106 294	102 443	0.0386	0.0151	0.0235	1 299 707	30 529
1984	0.0318	0.0051	0.0267	3 358 770	89 739	0.0289	0.0101	0.0188	5 212 666	98 127	0.0384	0.0164	0.0220	1 346 469	29 641
1985	0.0311	0.0055	0.0256	3 503 934	89 839	0.0299	0.0079	0.0220	5 336 337	117 480	0.0378	0.0145	0.0233	1 394 279	32 472
1986	0.0298	0.0063	0.0235	3 553 141	83 590	0.0307	0.0073	0.0235	5 456 753	127 982	0.0377	0.0127	0.0251	1 439 110	36 062
1987	0.0295	0.0059	0.0235	3 644 254	85 804	0.0312	0.0055	0.0257	5 632 094	144 537	0.0388	0.0178	0.0210	1 487 895	31 227
1988	0.0295	0.0067	0.0228	3 723 578	84 819	0.0293	0.0052	0.0241	5 801 174	139 911	0.0370	0.0226	0.0144	1 545 010	22 217
1989	0.0297	0.0064	0.0233	3 441 058	80 133	0.0280	0.0021	0.0259	5 999 477	155 299	0.0356	0.0210	0.0146	1 599 996	23 321
1990	0.0298	0.0064	0.0233	3 269 590	76 266	0.0296	−0.0031	0.0327	6 221 301	203 405	0.0355	0.0176	0.0179	1 641 181	29 423
1991	0.0309	0.0055	0.0254	2 990 844	75 903	0.0305	−0.0022	0.0326	6 417 480	209 463	0.0364	0.0208	0.0155	1 944 518	30 235
1992	0.0327	0.0054	0.0273	3 457 814	94 530	0.0285	0.0021	0.0264	6 491 160	171 475	0.0381	0.0197	0.0184	1 711 186	31 470
1993	0.0335	0.0049	0.0286	3 944 041	112 913	0.0288	0.0041	0.0247	6 617 141	163 414	0.0405	0.0213	0.0191	1 756 962	33 581
1994	0.0345	0.0060	0.0285	4 094 298	116 657	0.0288	0.0097	0.0191	6 668 240	127 540	0.0407	0.0232	0.0175	1 809 270	31 609
1995						0.0282	0.0181	0.0101	6 730 480	68 034	0.0400	0.0251	0.0149	1 861 386	27 752
1996						0.0341	0.0101	0.0240	6 803 898	163 222	0.0404	0.0232	0.0172	1 908 998	32 867
1997						0.0350	0.0109	0.0241	6 852 569	165 486	0.0411	0.0243	0.0168	1 958 562	33 000

Table 14.5 Scarcity ratios for KN

Year	KN/KM (DE)	KN_C/Y (DE)	P_{KN}/P_{KM} (DE)	KN/KM (SE)	KN_C/Y (SE)	P_{KN}/P_{KM} (SE)	KN/KM (AU)	P_{KN}/P_{KM} (AU)	KN_C/Y (AU)
1969	0.0969	0.0132	0.5212	1.0549	0.0481	0.7131	0.9993	0.4062	−0.0310
1970	0.1091	0.0133	0.4976	1.0580	0.0505	0.6882	0.9993	0.4363	−0.0057
1971	0.1104	0.0138	0.5301	1.0266	0.1067	0.9397	0.9993	0.4433	−0.0016
1972	0.1112	0.0140	0.5434	0.9964	0.0973	0.8925	0.9993	0.4130	−0.0197
1973	0.1152	0.0141	0.5497	1.0478	0.0794	0.7998	0.9993	0.4549	−0.0002
1974	0.1208	0.0147	0.5816	1.0449	0.0723	0.7219	0.9993	0.5740	0.0681
1975	0.2075	0.0262	0.6410	1.0382	0.0743	0.7305	0.9978	0.6725	0.0921
1976	0.2048	0.0252	0.6120	1.0050	0.1040	0.8605	0.9976	0.6423	0.0844
1977	0.2009	0.0249	0.6220	0.9895	0.1490	1.1518	0.9966	0.6853	0.1056
1978	0.1925	0.0238	0.6114	0.9748	0.1287	0.9735	0.9962	0.6056	0.0708
1979	0.1693	0.0214	0.6181	0.9943	0.1047	0.7920	0.9968	0.6002	0.0687
1980	0.1710	0.0229	0.6967	0.9994	0.1032	0.7744	0.9966	0.6425	0.0869
1981	0.2291	0.0326	0.7710	0.9991	0.1221	0.8718	0.9969	0.6928	0.1081
1982	0.3035	0.0459	0.8259	0.9986	0.0978	0.7297	0.9958	0.8586	0.1620
1983	0.3448	0.0527	0.8050	0.9985	0.1027	0.7571	0.9922	0.7617	0.1187

cont.

Table 14.5   Scarcity ratios for KN

Year	KN/KM (DE)	KN$_C$/Y (DE)	P$_{KN}$/P$_{KM}$ (DE)	KN/KM (SE)	KN$_C$/Y (SE)	P$_{KN}$/P$_{KM}$ (SE)	KN/KM (AU)	P$_{KN}$/P$_{KM}$ (AU)	KN$_C$/Y (AU)
1984	0.3447	0.0511	0.7668	0.9989	0.0942	0.7447	0.9941	0.7418	0.1093
1985	0.3513	0.0502	0.7366	0.9991	0.1108	0.8146	0.9949	0.7323	0.1155
1986	0.3479	0.0455	0.6755	0.9992	0.1177	0.8610	0.9949	0.7443	0.1260
1987	0.3485	0.0460	0.6871	0.9996	0.1289	0.9174	0.9953	0.7029	0.1034
1988	0.3476	0.0438	0.6510	0.9997	0.1224	0.8622	0.9963	0.6278	0.0704
1989	0.3131	0.0399	0.6360	0.9998	0.1330	0.8984	0.9970	0.6234	0.0716
1990	0.2893	0.0359	0.6069	0.9997	0.1722	1.1537	0.9957	0.6697	0.0913
1991	0.2569	0.0341	0.6309	0.9991	0.1795	1.1932	1.1526	0.6881	0.0934
1992	0.2888	0.0418	0.7059	0.9871	0.1494	0.9782	0.9908	0.6916	0.0940
1993	0.3223	0.0512	0.7823	0.9928	0.1467	0.9141	0.9923	0.7124	0.0952
1994	0.3278	0.0516	0.7565	0.9933	0.1100	0.7885	0.9929	0.7016	0.0851
1995				0.9936	0.0559	0.6401	0.9945	0.6742	0.0715
1996				0.9940	0.1320	0.9607	0.9943	0.7049	0.0821
1997				0.9942	0.1312	0.9703	0.9943	0.7046	0.0793

Table 14.6 Efficiency ratios

Year	KN/KM (DE)	KN/KM (SE)	KN/KM (AU)	ΔKM/Y (DE)	ΔKM/Y (SE)	ΔKM/ (AU)	KN/ΔKM (DE)	KN/ΔKM (SE)	KN/ΔKM (AU)
1969	0.0969	1.0549	0.9993	0.212	0.191	0.233	2.250	23.018	18.963
1970	0.1091	1.0580	0.9993	0.358	0.168	0.127	1.421	26.211	34.187
1971	0.1104	1.0266	0.9993	0.221	0.176	0.211	2.337	24.727	20.706
1972	0.1112	0.9964	0.9993	0.207	0.164	0.207	2.579	27.863	21.238
1973	0.1152	1.0478	0.9993	0.175	0.167	0.180	3.336	27.684	25.273
1974	0.1208	1.0449	0.9993	0.161	0.151	0.186	6.558	30.744	24.909
1975	0.2075	1.0382	0.9978	0.158	0.139	0.184	6.443	32.958	25.524
1976	0.2048	1.0050	0.9976	0.159	0.135	0.173	6.309	35.248	27.934
1977	0.2009	0.9895	0.9966	0.159	0.114	0.176	6.041	41.520	27.104
1978	0.1925	0.9748	0.9962	0.166	0.114	0.172	5.082	41.813	28.219
1979	0.1693	0.9943	0.9968	0.168	0.119	0.186	5.219	40.739	26.264
1980	0.1710	0.9994	0.9966	0.154	0.145	0.193	7.908	34.279	25.811
1981	0.2291	0.9991	0.9969	0.140	0.124	0.159	11.930	41.041	33.064
1982	0.3035	0.9986	0.9958	0.143	0.117	0.153	13.397	43.678	33.103
1983	0.3448	0.9985	0.9922	0.135	0.100	0.164	14.221	49.834	30.277

cont.

Table 14.6   Efficiency ratios (cont.)

Year	KN/KM (DE)	KN/KM (SE)	KN/KM (AU)	ΔKM/Y (DE)	ΔKM/Y (SE)	ΔKM/ (AU)	KN/ΔKM (DE)	KN/ΔKM (SE)	KN/ΔKM (AU)
1984	0.3447	0.9989	0.9941	0.129	0.116	0.167	15.190	43.522	29.664
1985	0.3513	0.9991	0.9949	0.130	0.110	0.157	14.898	45.551	31.951
1986	0.3479	0.9992	0.9949	0.130	0.155	0.161	15.035	32.500	30.652
1987	0.3485	0.9996	0.9953	0.132	0.147	0.177	14.522	34.410	27.703
1988	0.3476	0.9997	0.9963	0.138	0.170	0.166	12.439	30.317	29.628
1989	0.3131	0.9998	0.9970	0.148	0.188	0.135	10.439	27.966	37.664
1990	0.2893	0.9997	0.9957	0.153	0.172	0.120	8.790	32.054	50.184
1991	0.2569	0.9991	1.1526	0.146	0.133	0.119	10.498	42.518	42.859
1992	0.2888	0.9871	0.9908	0.120	0.080	0.124	14.950	74.484	40.221
1993	0.3223	0.9928	0.9923	0.113	0.042	0.139	16.010	138.017	35.159
1994	0.3278	0.9933	0.9929		0.050	0.128		111.395	37.523
1995		0.9936	0.9945		0.058	0.121		95.662	39.565
1996		0.9940	0.9943		0.038	0.120		143.018	39.351
1997		0.9942	0.9943						

Notes

1 Italics—my emphasis.
2 The explanation for this is given in most texts. Put simply, marginal costs are usually U-shaped, and in the descending region of marginal costs, average costs exceed marginal costs. Hence, responding to price changes in this region can result in only losses.
3 For some other authors such as Frank (2000), non-excludability is a characteristic of public goods. That is, it is either very expensive or impossible to prevent people from using a public good.
4 This chapter draws heavily on Thampapillai and Quah (2002).
5 Although this assumption can be restrictive, the framework presented here provides a basis for valuation and extension to consider the role of other non-pecuniary factors. However, given the limitations of data we persist with this assumption on the grounds that KN is the dominant component of most non-pecuniary factors.

References

Ahmad, Y.J., El Serafy, S., and Lutz, E. (eds) (1989), *Environmental Accounting for Sustainable Development*, World Bank, Washington, DC.

Anderson, J.R. and Thampapillai, D.J. (1990), *Soil Conservation in Developing Countries: Project and Policy Intervention*, Paper Number 8, Policy and Research Series, Agriculture and Rural Development Department, World Bank, Washington, DC.

Ayres, R.U. (1998), 'Eco-thermodynamics: Economics and the Second Law', *Ecological-Economics*, 26(2): 189–209.

—— (1999), 'The Second Law, the Fourth Law, recycling and limits to growth', *Ecological-Economics*, 29(3): 473–83.

Barnett, H. and Morse C. (1963), *Scarcity and Growth: The Economics of Natural Resource Availability*, Johns Hopkins University Press, Baltimore, Md.

Bateman, I.J., Munro, A., Rhodes, B., Starmer, C., and Sugden, R. (1997), 'A test of the theory of reference-dependent preferences', *Quarterly Journal of Economics*, 112(2): 479–505.

Baumol, W.J. (1977), *Economic Theory and Operations Analysis*, Prentice-Hall, Englewood Cliffs, NJ.

Baumol, W.J. and Blinder, A.S. (1988), *Economics: Principles and Policy*, Harcourt Brace Jovanovich, San Diego, Calif.

Bell, F.C. (1992), 'Prospects for Aquaculture in NSW', *Report 2—Developments in Aquaculture in NSW and their Potential Impacts*, Total Environment Centre, Sydney.

Blaug, M. (1958), *Ricardian Economics: A Historical Study*, Greenwood Press, New Haven, Conn.

—— (1986), *Great Economists Before Keynes: An Introduction to the Lives and Works of One Hundred Great Economists of the Past*, Wheatsheaf, Brighton, UK.

Blyth, M.J. and Kirby, M.G. (1985), 'The impact of government policy on land degradation in the rural sector', in A.J. Jakeman, D.G. Day, and A.K. Dragun (eds), *Policies for Environmental Quality Control*, CRES Monograph 15, Australian National University, Canberra.

Böjo, J., Mäler, K-G., and Unemo, L. (1990), *Environment and Development: An Economic Approach*, Kluwer Academic Publishers, Dordrecht.

Boulding, K.E. (1993), *The Structure of a Modern Economy: The United States, 1929–89*, New York University Press, New York.

Brandon, K.E. and Brandon, C. (1992), 'Linking environment to development: Problems and possibilities: Introduction', *World Development*, 20(4): 477–9.

Brundtland Commission (1987), *Food 2000: Global Policies for Sustainable Agriculture*, Zed Books, London.

Bunce, A.C. (1942), *The Economics of Soil Conservation*, University of Nebraska Press, Lincoln, Neb.

Burley, P. and Foster, J. (eds) (1994), *Economics and Thermodynamics: New Perspectives on Economic Analysis*, Kluwer Academic, Dordrecht and Boston.

Carson, R.T., Mitchell, R.C., Hanemann, W.M., Kopp, R.J., Presser, S., and Ruud, P.A. (1992), *A Contingent Valuation Study of Lost Passive Use Values Resulting from the Exxon Valdez Oil Spill: A Report to the Attorney-General of the State of Alaska*, November 1992.

Ciriacy-Wantrup, S. von (1938), 'Soil conservation in European farm management', *Journal of Farm Economics*, 20(1): 86–101.

Clawson, M. and Knetsch, J.L. (1966), *Economics of Outdoor Recreation*, Johns Hopkins University Press, Baltimore, Md.

Coase, R. (1960), 'The problem of social cost', *Journal of Law and Economics*, 3(1): 1–44.

Coomber, N.H. and Biswas, A.K. (1973), *Evaluation of Environmental Intangibles*, Genera Press, New York.

Costanza, R. (ed.) (1991), *Ecological Economics: Science and Management of Sustainability*, Columbia University Press, New York.

Cropper, M.L. and Oates, W.E. (1992), 'Environmental economics: A survey', *Journal of Economic Literature*, 30(2): 675–740.

Cruz, W. and Repetto, R. (1992), *Environmental Effects of Stabilization and Structural Adjustment Programs: The Philippines Case*, World Resources Institute, Washington, DC.

Cummings, R.G. and Harrison, G.W. (1992), 'Was the Ohio Court informed in their assessment of the accuracy of the contingent valuation method?', Paper B-92-07, College of Business Administration, University of South Carolina, Columbia, SC.

Daly, H.E. (1987), 'The economic growth debate: What some economists have learned but many have not,' *Journal of Environmental Economics and Management*, 14(4): 323–36.

—— (1989), 'Toward a measure of sustainable net national product', in Y.J. Ahmad, S. El Serafy, and E. Lutz (eds), *Environmental Accounting for Sustainable Development*, World Bank, Washington, DC.

—— (1991), 'Towards an environmental macroeconomics', *Land Economics*, 67(2): 255–9.

—— (1992), 'Is the Entropy Law relevant to the economics of natural resource scarcity? Yes, of course it is! Comment', *Journal of Environmental Economics and Management*, 23(1): 91–5.

—— (1996), *Beyond Growth: The Economics of Sustainable Development*, Beacon Press, Boston, Mass.

—— (1997a), 'Sustainable growth: An impossibility theorem', *Development*, 40(1): 121–5.

—— (1997b), 'Georgescu-Roegen versus Solow/Stiglitz', *Ecological-Economics*, 22(3): 261–6.

Daly, H.E. and Cobb, J.B. (1989), *For the Common Good: Redirecting the Economy Towards Community, the Environment, and a Sustainable Future*, Beacon Press, Boston, Mass.

Dasgupta, P. and Heal, G.M. (1979), *Economic Theory of Exhaustible Resources*, Cambridge University Press, Cambridge.

Dasgupta, P., Kriström, B., and Mäler, K-G. (1994), *Current Issues in Resource Accounting*, Report 105, Umeå, Department of Forest Economics, Swedish University of Agricultural Sciences, Uppsala, Sweden.

Dixon, J.A., Scura, L.A., Carpenter, R.A., and Sherman, P.B. (1994), *Economic Analysis of Environmental Impacts*, 2nd edn, Earthscan Publications, London.

Domar, E.D. (1946), 'Capital expansion, rate of growth and employment', *Econometrica*, 14(2): 137–47.

Dornbusch, R. and Fischer, S. (1994), *Macroeconomics*, McGraw-Hill, New York.

Environment Australia (1997), *Environmental Economics Round Table*, Proceedings, Department of the Environment, Canberra ACT.

—— (2001), 'Finance sector examines the triple bottom line', *Envirobusiness Update*, Issue 6.

Fisher, A.C. and Krutilla, J.V. (1985), 'Economics of nature preservation', in A.V. Kneese and J.L. Sweeney (eds), *Handbook of Natural Resource and Energy Economics*, North Holland, Amsterdam.

Fisher, I. (1904), 'Precedents for defining capital', *Quarterly Journal of Economics*, 18(3): 386–408.

Forrester, J.W. (1971), *World Dynamics*, Wright-Allen Press, Cambridge, Mass.

Franciosi, R., Kujal, P., Michelitsch, R., Smith, V., and Deng, G. (1996), 'Experimental tests of the endowment effect', *Journal of Economic Behaviour and Organisation*, 30(2): 213–26.

Frank, R. H. (2000), *Microeconomics and Behavior*, McGraw-Hill, New York.

French, D. (1986), 'Confronting an unsolvable problem: Deforestation in Malawi', *World Development*, 14(4): 531–40.

Frykblom, P. (1997), 'Hypothetical question modes and real willingness to pay', *Journal of Environmental Economics and Management*, 34(3): 275–7.

Georgescu-Roegen, N. (1971), *The Entropy Law and the Economic Process*, Harvard University Press, Cambridge, Mass.

Girma, M. (1992), 'Macropolicy and the environment: A framework for analysis', *World Development*, 20(4): 531–40.

Glahe, F.R. (1977), *Macroeconomics: Theory and Policy*, Harcourt Brace Jovanovich, New York.

Gray, L.C. (1914), 'Rent under the assumption of exhaustibility', *Quarterly Journal of Economics*, 28(3): 466–89.

Greene, W.H. (1993), *Econometric Analysis*, 2nd edn, Macmillan, New York.

Greig, P.J. and Devonshire, P.G. (1981), 'Tree removals and saline seepage in Victorian catchments: Some hydrologic and economic results', *Australian Journal of Agricultural Economics*, 25:134–48.

Hahn, R.W. and Hester, G.L. (1989), 'Marketable permits: Lessons for theory and practice', *Ecology Law Quarterly*, 16(2).

Hamilton, C., Hundloe, T., and Quiggin, J. (1997), *Ecological Tax Reform in Australia: Using Taxes and Public Spending to Protect the Environment Without Hurting the Economy*, Discussion Paper 10, Australia Institute, Canberra.

Harrod, R.F. (1939), 'An essay in dynamic theory', *The Economic Journal*, 49(193): 14–33.

―― (1948), *Towards a Dynamic Economics*, Macmillan, London.

Hartwick, J. (1977), 'Intergenerational equity and the investing of rents from exhaustible resources', *American Economic Review*, 66: 972–4.

―― (1978), 'Investing returns from depleting renewable resource stocks and intergenerational equity', *Economic Letters*, 1(1): 85–8.

―― (1990), 'Natural resources, national accounting and economic depreciation', *Journal of Public Economics*, 43(3): 291–304.

―― (1993), 'Forestry economics, deforestation, and national accounting', in E. Lutz. (ed.), *Toward Improved Accounting for the Environment*, UNSTAT–World Bank Symposium, World Bank, Washington, DC.

Hartwick, J. and Olewiler, N. (1986), *Economics of Natural Resource Use*, Harper & Row, New York.

Hecht, S. (1985), 'Environment, development and politics: Capital accumulation and the livestock sector in Eastern Amazonia', *World Development*, 13(6): 663–84.

Hirshleifer, J. (1988), *Price Theory and Applications*, Prentice-Hall, Englewood Cliffs, NJ.

Hotelling, H. (1931), 'The economics of exhaustible resources', *Journal of Political Economy*, 39(2): 137–75.

―― (1949), letter quoted in *Economics of Public Recreation: An Economic Study of the Monetary Evaluation of Recreation in the National Parks Service*, National Parks Service, United States Department of the Interior.

Hundloe, T. (1997), 'Achieving environmental objectives by the use of economic instruments: Fisheries', *Environmental Economics Round Table Proceedings*, Environment Australia, Canberra.

Jackson, D. (1989), *The Australian Economy*, Macmillan, Melbourne.

James, D. (1997), *Environmental Incentives: Australian Experience with Economic Instruments for Environmental Management*, Environmental Economics Research Paper No. 5, Environment Australia, Canberra.

Jevons, W.S. (1866), *An Inquiry Concerning the Progress of the Nation, and the Probable Exhaustion of Our Coal-Mines*, Macmillan & Co., London.

Jorgenson, D.W. (1967), 'The theory of investment behaviour', in R. Ferber (ed.), *Determinants of Investment Behaviour*, Columbia University Press, New York.

Kahneman, D., Knetsch, J.L., and Thaler, R.H. (1990), 'Experimental tests of the endowment effect and the Coase theorem', *The Journal of Political Economy*, 98(6), 1325–48.

Katz, M.L. and Rosen, H.S. (1991), *Microeconomics*, Irwin, Burr Ridge, Ill.

Keynes, J. (1936), *The General Theory of Employment, Interest and Money*, Macmillan, London.

Kneese, A.V., Ayres, R.U., and D'Arge, R.C. (1970), *Economics and the Environment: A Material Balance Approach*, Resources for the Future, Washington, DC.

Knetsch, J.L. (1963), 'Outdoor recreation demands and benefits', *Land Economics*, 39: 387–96.

—— (1964), 'Economics of including recreation as a purpose of eastern water projects', *Journal of Farm Economics*, 46: 1148–57.

—— (1989), 'The endowment effect and evidence of nonreversible indifference curves', *American Economic Review*, 79(5): 1277–84.

—— (1994), 'Environmental valuation: Some problems of wrong questions and misleading answers', *Environmental Values*, 3(4): 351–68.

—— (1995), 'Asymmetric valuation of gains and losses and preference order assumptions', *Economic Inquiry*, 38(1): 138–41.

Knetsch, J.L. and Sinden, J.A. (1984), 'Willingness to pay and compensation demanded: Experimental evidence of an unexpected disparity in measures of value', *Quarterly Journal of Economics*, 99(3): 507–21.

Krutilla, J.V. and Cicchetti, C.J. (1972), 'Evaluating benefits of environmental resources with special application to the Hells Canyon', *Natural Resources Journal*, 12: 1–29.

Lecomber, R. (1979), *The Economics of Natural Resources*, Macmillan, London.

Lutz, E. (ed.) (1993), *Toward Improved Accounting for the Environment*, UNSTAT–World Bank Symposium, World Bank, Washington, DC.

Lutz, E. and Peskin, H. (1993), 'A survey of resource and environmental accounting approaches in industrialised countries', in E. Lutz (ed.), *Toward Improved*

Accounting for the Environment, UNSTAT–World Bank Symposium, World Bank, Washington DC.

McInerney, J. (1981), 'The simple analytics of natural resource economics', in J.A. Butlin (ed.), *Economics and Resources Policy*, Longman, London.

Maddala, G. (1992), *Introduction to Econometrics*, 2nd edn, Macmillan, New York.

Mäler, K-G. (1974), *Environmental Economics: A Theoretical Inquiry*, Johns Hopkins University Press, Baltimore, Md.

—— (1991), 'National accounting and environmental resources', *Environmental and Resource Economics*, 1(1): 1–15.

Malthus, T. (1798), *An Essay on the Principle of Population, as it Affects the Future Improvement of Society with Remarks on the Speculations of Mr. Godwin, M. Condorcet, and Other Writers*, J. Johnson, St Paul's Church-Yard, London.

Mankiw, N.G. (1998), *Principles of Macroeconomics*, Dryden Press, Orlando, Fla.

Marshall, A. (1891), *Principles of Economics*, Macmillan, London.

Marxsen, C.S. (1992), 'Towards an environmental macroeconomics: Comment', *Land Economics*, 68(2): 241–3.

Meadows, D.H., Meadows, D.L., Randers, J., and Behrens, W.W. (1972), *The Limits to Growth: A Report of the Club of Rome's Project on the Predicament of Mankind*, Earth Island, Universe Books, New York.

Mendelsohn, R. (1992), 'Measuring hazardous waste damages with panel models', *Journal of Environmental Economics and Management*, 22(3): 259–71.

Mill, J.S. (1848), *Principles of Political Economy, with Some of Their Applications to Social Philosophy*, John W. Parker, West Strand, London.

Nadkarni, M.V. (1987), 'Agricultural development and ecology—An economist's view', *Indian Journal of Agricultural Economics*, 42(3): 360–75.

Newcombe, K.J. (1989), 'An economic justification for rural afforestation: The case of Ethiopia', in G. Schramm and J.J. Warford (eds), *Environmental Management and Economic Development*, Baltimore: Johns Hopkins University Press for the World Bank.

Neill, R.H., Cummings, R.G., Ganderton, P.T, Harrison, G.W., and McGuckin, T. (1993), 'Hypothetical surveys and real commitments', Paper B-93-01, College of Business Administration, University of South Carolina, Columbia, SC.

Nordhaus, W.D. (1973), 'World dynamics: Measurement without data', *Economic Journal*, 83(332): 1156–83.

OECD (1997), *Flows and Stocks of Fixed Capital*, OECD, Paris.

—— (1998), *National Accounts: Volume II*, OECD, Paris.

—— (1999), *National Accounts: Volume I*, OECD, Paris.

Osiatynski, J. (ed.) (1990), *Collected Works of Michal Kalecki*, volumes I–V, Clarendon Press, Oxford.

Pearce, D.W. and Turner, K. (1990), *Environmental and Natural Resource Economics*, Wheatsheaf and Harvester, New York.

Pigou, A. (1920), *The Economics of Welfare*, Macmillan, London.

Repetto, R. (1986), 'Soil loss and population pressure on Java', *Ambio*, 15(1): 14–18.

Repetto, R. and Gillis, M. (eds) (1988), *Public Policy and the Misuse of Forest Resources*, Cambridge University Press, Cambridge.

Repetto, R. and Magrath, W.B. (1989), *Wasting Assets: Natural Resources in National Income Accounts*, World Resources Institute, Washington, DC.

Ricardo, D. (1817), *On the Principles of Political Economy and Taxation*, John Murray, London.

Samuelson, P.A. (1948), *Economics*, McGraw-Hill, New York, 1970 edn.

Samuelson, P.A. and Nordhaus, W. (1990), *Economics*, McGraw-Hill, New York.

Schickele, R. (1935), 'Economic implications of erosion control in the Corn Belt', *Journal of Farm Economics*, 17(3): 433–48.

Scott, A. (1954), 'Conservation policy and capital theory', *Canadian Journal of Economics and Political Science*, 20(4): 504–13.

Sinden J.A. (1967), 'The evaluation of extra market benefits: A review', *World Agricultural Economics and Rural Sociology Abstracts*, December.

—— (1973), 'Utility analysis in the valuation of extra-market benefits with particular reference to water based recreation', *Water Resources Institute Bulletin*, WRRI-17, Corvallis, Ore.

—— (1974), 'A utility approach to the valuation of recreational and aesthetic experiences', *American Journal of Agricultural Economics*, 56:61–72.

Sinden, J.A. (ed.) (1972), *The Natural Resources of Australia: Problems and Prospects for Development*, McGraw-Hill, Sydney.

Sinden, J.A. and King, D.A. (1988), Land condition, crop productivity, and the adoption of soil conservation measures, paper presented to the Australian Agricultural Economics Society Conference, Melbourne, February 1988.

Sinden, J.A. and Worrell, A.C. (1979), *Unpriced Values—Decisions without Market Prices*, Wiley, New York.

Sinden, J.A. and Wyckoff, J.B. (1976), 'Indifference mapping—An empirical methodology for economic valuation of the environment', *Regional Science and Urban Economics*, 6:81–103.

Sinden, J.A. and Yapp, T. (1987), The opportunity cost of land degradation in New South Wales: A case study, paper presented to the Australian Agricultural Economics Society Conference, Adelaide, February 1987.

Smith, A. (1776), *An Inquiry into the Nature and Causes of the Wealth of Nations*, Strahan & Cadell, London.

Solow, R. (1956), 'A contribution to the theory of economic growth', *Quarterly Journal of Economics*, 70(1): 65–94.

Solow, R.M. (1974), 'Intergenerational equity and exhaustible resources', *Review of Economic Studies: Symposium on the Economics of Exhaustible Resources*, Volume 41 (Symposium on the Economics of Exhaustible Resources): 29–45.

—— (1975), 'Reswitching: Brief comments', *Quarterly Journal of Economics*, 89(1): 48–52.

—— (1986), 'On the intergenerational allocation of natural resources', *Scandinavian Journal of Economics*, 88(1): 141–9.

—— (1992), 'An almost practical step toward sustainability', Invited Lecture on the Occasion of the Fortieth Anniversary of Resources for the Future, Resources for the Future, Washington, DC, October 1992.

Thampapillai, D.J. (1985), 'Trade-offs for conflicting social objectives in the extraction of finite energy resources', *International Journal of Energy Research*, 9:179–92.

—— (1988), 'The value of natural environments in the extraction of finite energy resources: A method of valuation', *International Journal of Energy Resources*, 12:527–38.

—— (1992), *Environmental Economics*, Oxford University Press, Melbourne.

—— (1993), *Environmental Macroeconomics: Some Conceptual Considerations*, Småsskriftsserien nr 73, Department of Economics, Swedish University of Agricultural Sciences, Uppsala, Sweden.

—— (1995a), 'Environment and the macroeconomy: Some conceptual considerations within a Keynesian framework', *Australasian Journal of Regional Studies*, December.

—— (1995b), 'Environmental macroeconomics: Towards filling an empty box', *Indian Economic Journal*, April–June.

—— (2000), 'Willingness to pay and willingness to accept: A simple conceptual exposition', *Applied Economics Letters*, 7(8):509–11.

Thampapillai, D.J., Maleka, P.T., and Milimo, J. (1992), 'Quantification of the trade-offs between environment, employment, income and food security', *Working Paper No. 229*, International Labour Organisation, Geneva, Switzerland.

Thampapillai, D.J. and Quah, E., 'Environmental Valuation', chapter 4 in Quah, E., and Tan, K.C., *Siting Environmentally Unwanted Facilities: Risks, Trade-offs and Choices*, Edward Elgar, London, 2002.

Thampapillai, D.J. and Öhlmer, B. (2000), *Environmental Economics for Business Management*, Department of Economics, Swedish University of Agricultural Sciences, Uppsala, Sweden.

Thampapillai, D.J. and Uhlin, H-E. (1994), *On the Measurement of Environmental Depreciation in Macroeconomic Analysis*, Working Paper 12, Department of Economics, Swedish University of Agricultural Sciences, Uppsala, Sweden.

—— (1996), 'Sustainable income: Extending some Tisdell considerations to macroeconomic analyses', *International Journal of Social Economics*, 23(4-5-6): 137–50.

—— (1997), 'Environmental capital and sustainable income: Basic concepts and empirical tests', *Cambridge Journal of Economics*, 21(3): 379–94.

Tisdell, C.A. (1991), *Economics of Environmental Conservation*, Elsevier, Amsterdam.

Todaro, M.P. (2000), *Economic Development*, Addison Wesley, New York.

Vuuren, W. van (1986), 'Soil erosion: The case of market intervention', *Canadian Journal of Agricultural Economics*, 33: 41–62.

Walker, D.J. (1982), 'A damage function to evaluate erosion control economics', *American Journal of Agricultural Economics*, 64: 690–8.

Walker, D.J. and Young, D.L. (1986), 'The effects of technical progress on erosion damage and economic incentives for soil conservation', *Land Economics*, 62: 83–93.

Wallace, N.W. (ed.) (1992), *Natural Resource Management—An Economic Perspective*, Australian Bureau of Agricultural and Resource Economics, AGPS, Canberra.

Water Resources Engineers Inc. (1970), *Wild River Method of Evaluation*, United States Department of the Interior, Office of Water Resources, Washington, DC.

Whittington, D., Okorafor, A., and Okore, A. (1990), 'Strategy for cost recovery in the rural water sector: A case study of Nuskka District, Anambra State, Nigeria', *Water Resources Research*, 26(9): 1899–1913.

World Bank (1992), *World Development Report*, Oxford University Press, New York.

—— (1994), *World Tables 1994*, Oxford University Press, New York.

Young, M.D. (1992), *Sustainable Investment and Resource Use Equity, Environmental Integrity and Economic Efficiency*, Man and the Biosphere series, volume 9, UNESCO and Parthenon Publishing, Carnforth, UK.

Index

Page numbers in *italics* refer to figures and tables.